Peterson
First Guide

to

BIRDS
of North America

Roger Tory Peterson

HOUGHTON MIFFLIN COMPANY
BOSTON   NEW YORK

PARTS OF A BIRD

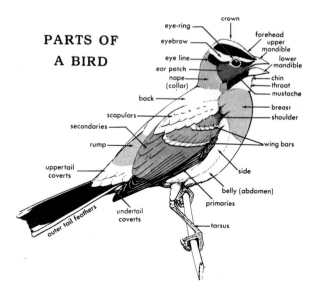

PETERSON FIRST GUIDES, PETERSON FIELD GUIDES, and
PETERSON FIELD GUIDE SERIES are registered
trademarks of Houghton Mifflin Company.

Selected illustrations reproduced from
A Field Guide to the Birds, 4th edition, copyright
© 1980 by Roger Tory Peterson.

Library of Congress Cataloging-in-Publication Data

Peterson, Roger Tory, date
Peterson First guide to birds of North America

Includes index.
1. Birds—North America—Identification. I. Title.
II. Title: First Guide to birds of North America.
QL681.P46 1986 598.2973 85-27102

ISBN 0-395-90666-0

Printed in Italy

NIL 30 29 28 27 26 25 24

INTRODUCTION

In 1934, when birdwatching was beginning to emerge from the shotgun or specimen-tray era of ornithology, my *Field Guide to the Birds* first saw the light of day. This book was designed so that live birds could be readily identified at a distance by their "field marks" without resorting to the bird-in-hand characters that the early collectors relied upon. During the half century since my guide appeared the binocular has replaced the shotgun.

Earlier handbooks were too complex. Birds were described systematically from beak to tail; thus the Robin's description might read: "Yellow beak with dark tip, blackish head with a white spot above the eye and one below, black streaks on a white throat, gray back, lighter feather-edgings, etc." Only halfway down the page would the rusty breast be mentioned.

It seemed to me that there must be an easier route; and thus was born the "Peterson System," as it is now called. The essence of the system is *simplification,* not amplification. It is a visual system, based on patternistic drawings with arrows that pinpoint the key field marks.

Later editions added more illustrations showing more plumages. These met the growing needs of the average birder, but many beginners faced with so many choices were confused. Where to start? It is for this audience—anyone who perhaps recognizes Crows, Robins, House Sparrows, and maybe Chickadees and Blue Jays, but little else—that this primer has been prepared. It is a selection of the birds you are most likely to see during your first forays afield.

However, this is only an introduction to the delights of birding. The females of many species, dressed in drabber tones, are not shown extensively in this First Guide. Most female ducks are brown and speckled, so my advice to the beginner is to recognize them by the company they keep. The males are easy.

But it will not be long before you are ready for the full treatment. You will want to acquire a copy, either in flexible binding or hardcover, of *A Field Guide to the Birds* or its western counterpart, *A Field Guide to Western Birds*. These two books cover *all* the species normally found north of the Mexican border, and most of the different plumages. The maps, fairly large in the latest eastern guide (1980), will inform you where to expect each species.

Should you wish to add sound and movement to the text and pictures of the *Field Guide*, or if you are just an armchair birder, you will learn the tricks of identification more quickly if you acquire the videocassette *Watching Birds*, prepared by Houghton Mifflin. Aside from its instructional value, this hour-long presentation is a feast for the eyes and ears.

Some birders depend on their ears as much as their eyes; I do. Words may describe a bird's voice, but there is no substitute for the sounds themselves. With the collaboration of the Cornell Laboratory of Ornithology, we have prepared records and cassettes of actual songs recorded in the field. Arranged to accompany the Field Guides, they are entitled: *A Field Guide to Bird Songs of Eastern and Central North America* and *A Field Guide to Western Bird Songs*. Play them at home and enjoy them.

HOW TO IDENTIFY BIRDS

If you are a beginner, and I presume you are,
you should become familiar in a general way
with the illustrations in this pocket guide,
which are excerpted from *A Field Guide to the
Birds*. The birds are not arranged in systematic or phylogenetic order but are grouped
in 8 main visual categories:

(1) **Ducklike Swimmers**—Ducks and
ducklike birds, pp. 18–31.
(2) **Aerialists**—Gulls and gull-like birds, pp.
32–37.
(3) **Long-legged Waders**—Herons and
egrets, pp. 38–41.
(4) **Smaller Waders**—Plovers and sandpipers, pp. 42–47.
(5) **Fowl-like Birds**—Grouse, quail, etc., pp.
48–49.
(6) **Birds of Prey**—Hawks, owls, etc., pp.
50–57.
(7) **Nonpasserine Land Birds**—Doves,
cuckoos, swifts, etc., pp. 58–65.
(8) **Passerine (Perching) Birds**, pp. 66–126.

Within these groupings you will see, for
example, that ducks are unlike loons and
gulls are readily distinguishable from terns.
The needle-like bills of warblers separate them
from the sparrows, which have thick seedcracking bills. Note the following points:

WHAT IS THE BIRD'S SIZE?
Make a habit of comparing the size of a new
bird with that of some familiar bird—a House
Sparrow, Robin, Pigeon, or whatever, so that
you can say to yourself, "smaller than a
Robin; a little larger than a House Sparrow."
The measurements in this book are in inches
from bill-tip to tail-tip.

DOES IT SWIM? IS IT DUCKLIKE?

Does the bird sit low in the water like a loon (1), or high like a Moorhen or Coot (2)? If it is a duck, does it dive like a deep-water bay or sea duck (3), or does it dabble and up-end like a marsh duck such as the Mallard (4)?

WING PATTERNS

The basic wing patterns of ducks (shown below), shorebirds, and other water birds are very important. Notice whether the wings have patches (1) or stripes (2), are solidly colored (3), or have contrasting black tips (Snow Goose, Gannet, etc.).

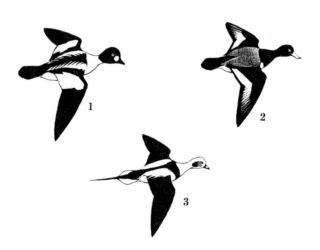

DOES IT WADE?

Is the bird large and long-legged like a heron, or small like a sandpiper? If one of the latter, does it probe the mud or pick at things? Does it teeter or bob?

HOW DOES IT FLY?

Does it undulate (dip up and down in the air) like a Flicker (1)? Does it fly straight and fast like a Mourning Dove (2)? Does it hover like a Kingfisher (3)? Does it glide or soar like a gull or a hawk?

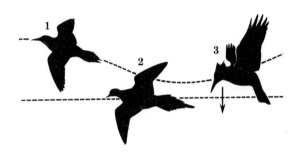

WHAT SHAPE ARE ITS WINGS?
Are they rounded like a Bobwhite's (left), or
sharply pointed like a Barn Swallow's (right)?

WHAT IS ITS GENERAL SHAPE?
Is it plump like a Starling (left), or slender like
a Cuckoo (right)?

WHAT SHAPE IS ITS TAIL?

Is it (1) Deeply forked, like a Barn Swallow's?
    (2) Square-tipped, like a Cliff Swallow's?
    (3) Notched, like a Tree Swallow's?
    (4) Rounded, like a Blue Jay's?
    (5) Pointed, like a Mourning Dove's?

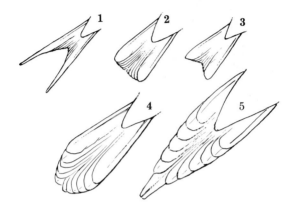

WHAT SHAPE IS ITS BILL?

Is it small and fine, like that of a warbler (1)?
Stout and short, adapted for seed cracking,
like that of a sparrow (2)? Dagger-shaped, like
that of a tern (3)? Or hook-tipped, for tearing
flesh, like the beak of a bird of prey (4)? Or
curved, like that of a creeper (p. 78)?

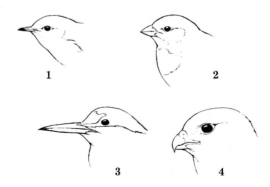

DOES IT CLIMB TREES?

If so, does it climb in spirals like a creeper (left)? In jerks like a woodpecker (center), using its tail as a brace? Or does it go down headfirst like a nuthatch (right)?

HOW DOES IT BEHAVE?

Does it cock its tail like a wren, or hold it down like a flycatcher? Does it wag its tail? Does it sit erect on an open perch, dart after an insect, and return, as a flycatcher does?

WHAT ARE ITS FIELD MARKS?

Some birds can be identified by color alone, but the most important aids are *field marks*, which are the "trademarks of nature." Note whether the breast is spotted, as in the Wood Thrush (1); streaked, as in the Brown Thrasher (2); or plain, as in a cuckoo (3).

TAIL PATTERNS

Does the tail have a "flash pattern"—a white band at the tip, as in the Kingbird (1)? White patches in the outer corners, as in the Towhee (2)? Or white sides, as in the Junco (3)?

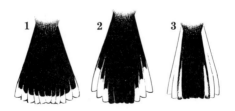

RUMP PATCHES

Does it have a light rump like a Cliff Swallow (1) or a Flicker (2)? Many of the shorebirds also have distinctive rump patches. So does the Harrier.

WING BARS

Do the wings have light wing bars or not? Their presence or absence is important in recognizing many warblers, vireos, and flycatchers. Wing bars may be single or double, bold or obscure.

CRESTS

Does the bird have a topknot or crest? Is it pronounced, like that of a Cardinal, a Waxwing, a Titmouse, or a Blue Jay? Or is it a slight crest, created by raising the crown feathers, as in a Crested Flycatcher or a White-crowned Sparrow?

EYE-STRIPES AND EYE-RINGS

Does the bird have a stripe above the eye, through the eye, or below it, or a combination of these stripes? Does it have a striped crown? A ring around the eye, or "spectacles?" A "mustache" stripe? These details are important in many small songbirds.

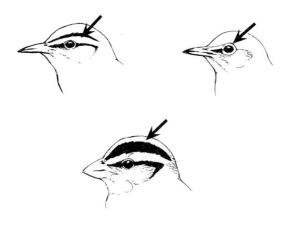

Roadside Silhouettes

1	MOURNING DOVE	12	ROBIN
2	HOUSE SPARROW	13	KILLDEER
3	GRACKLE	14	PHEASANT
4	STARLING	15	KINGFISHER
5	COWBIRD	16	PURPLE MARTIN
6	RED-WINGED	17	BARN SWALLOW
	BLACKBIRD	18	CLIFF SWALLOW
7	BLUE JAY	19	KESTREL
8	MOCKINGBIRD	20	FLICKER
9	SONG SPARROW	21	CARDINAL
10	BLUEBIRD	22	MEADOWLARK
11	NIGHTHAWK	23	KINGBIRD

Flight Silhouettes

1	BARN SWALLOW	13	RUFFED GROUSE
2	CLIFF SWALLOW	14	KESTREL
3	PURPLE MARTIN	15	GRACKLE
4	CHIMNEY SWIFT	16	RED-WINGED
5	STARLING		BLACKBIRD
6	BLUEBIRD	17	BLUE JAY
7	GOLDFINCH	18	MOURNING DOVE
8	HOUSE SPARROW	19	ROBIN
9	KINGFISHER	20	KILLDEER
10	FLICKER	21	WOODCOCK
11	MEADOWLARK	22	PHEASANT
12	BOBWHITE		

DUCKLIKE SWIMMERS

LOONS AND GREBES. Loons are longish swimming birds with daggerlike bills. They may dive or submerge. Of the 4 species in the world, all are found in North America. **Grebes** are mostly smaller than loons, with a tail-less look. They have lobed toes. There are 20 species in the world; 5 in North America.

COMMON LOON   28–36″

This avian submarine, identified by its longish profile, *checkerboard* pattern, and *daggerlike bill,* spends the summer on forested lakes of Canada and the "golden ponds" of our border states, where its yodeling cries lend mystery to the night. In winter on salt water of both coasts its checkerboard pattern is replaced by gray.

HORNED GREBE   12–15″

A small diver with a slender pointed bill. It is identified by its *golden ear tufts* and *chestnut neck*; in winter, it has white cheeks and a black cap. Found on northern lakes in summer, coastal waters in winter. The **Eared Grebe** (not shown) of the West has a *black* neck in summer, a "dirtier" face and neck in winter.

PIED-BILLED GREBE   13″

This pond-loving grebe is chubby, with a puffy white stern. In summer—not in winter—it has a *black bib* and a *black ring* on its chicken-like bill. Continentwide in summer, it retreats to southern waters and coastal bays in winter. Its voice is a cuckoo-like *kuk-kuk-cow-cow-cow-cow-cowp-cowp.*

WESTERN GREBE   25″

This elegant swimmer with its *swanlike neck* is seen around fishing wharfs and bays along our West Coast except during the warmer months, when it retreats to inland waters to nest.

COMMON LOON

winter

summer

summer

winter

PIED-BILLED GREBE

summer

HORNED GREBE

winter

WESTERN GREBE

display

CORMORANTS are large, blackish water birds that live mainly along the coast. They fish for a living. These somber birds stand erect on rocks or posts and often strike a "spread-eagle" pose. There are 30 species in the world; 6 in North America.

DOUBLE-CRESTED CORMORANT   33″

Of the 6 species of cormorants in North America, this is the only one found on both coasts, where it nests colonially on rocky islands. On lakes it may nest in trees. These big silent birds fly in lines or wedges like geese and swim low in the water like loons, but carry their hook-tipped bills tilted up at an angle. At close range the *yellow throat pouch* is apparent. If you see a cormorant on an inland lake it is certainly this species, but along the coasts there are other possibilities. To separate them you must study the *Field Guides.*

ANHINGAS or "Darters" are more restricted to freshwater swamps, although in some localities cormorants and anhingas can be seen in the same environment. Of the 4 species in the world, we have one.

AMERICAN ANHINGA   34″

The "Water Turkey," as it is sometimes called, is a bit like a cormorant, often perching upright on a snag and holding its wings out to dry, but it has a much *slimmer, snakier* look and a *longer tail.* Its wings have large *silvery* patches and the pointed bill lacks the hooked tip of the cormorant. The Anhinga lies low in the water like a cormorant but may submerge its body completely, swimming with only its head emergent; then it suggests a snake. Its domain is the swamps and waterways of the Southeast.

immature

adult

DOUBLE-CRESTED
CORMORANT

male

female

AMERICAN
ANHINGA

WATERFOWL. This cosmopolitan family, which includes ducks, geese, and swans, numbers 145 species in the world; in North America there are 44, plus 13 visitors or strays from abroad. **Geese** (6 in North America) are larger, heavier-bodied, and longer-necked than ducks. **Swans** (3 in North America) are huge, all-white, and longer-necked than geese.

TUNDRA (WHISTLING) SWAN   53″

This, our common native swan, with a wingspan of 6 to 7 feet, was formerly called the "Whistling" Swan. It can be told from the semidomestic park species, the **Mute Swan** (not shown), by its bill, which is *black* instead of orange. After nesting in the high Arctic, long skeins of Tundra Swans travel overland by way of the Great Lakes to their wintering grounds on the bays of the mid-Atlantic coast, or by way of certain western lakes to the central valleys of California. A larger relative, the rare **Trumpeter Swan** (not shown), lives on lakes in the wooded northwestern part of the continent.

SNOW GOOSE   25–38″

Snow Geese are readily separated from swans by their *black* primary wing feathers. After breeding in the Arctic, huge flocks travel across the continent to spend the winter in the marshes of the mid-Atlantic and Gulf coasts and the valleys of central California.

CANADA GOOSE   25–43″

Canada Geese come in various sizes but always have black neck stockings and *white chin straps.* After breeding across the upper half of the continent, some Canada Geese migrate as far south as Mexico, traveling in line or wedge formation.

TUNDRA
(WHISTLING)
SWAN

SNOW
GOOSE

CANADA GOOSE

Marsh Ducks. These surface feeders of ponds, marshes, and creeks feed by dabbling and tipping up. They do not dive. Males are distinctive, females less so—they may be known by the company they keep. Most marsh ducks are found continentwide, nesting in the northern interior, and migrating to coastal marshes in winter. Except for the Mallard, only males are shown here. Master these. Later, with either the eastern or western *Field Guide* you may learn the more subtle females, but to repeat—they are usually in the company of the well-marked males.

MALLARD   20–28″

The best-known duck; the ancestor of most barnyard quackers. Males have a *green-glossed head* and a ruddy chest, separated by a *white neck-ring.* Females are brown with a *yellowish bill* and a *whitish tail.* The females are the ones that quack.

AMERICAN BLACK DUCK   21–25″

Dark-bodied with a paler head and, in flight, *flashing white wing linings.* The Black Duck is not a western duck.

COMMON PINTAIL   26–30″

Slim, elegant. Note the *white neck stripe* and *needle-pointed tail.*

NORTHERN SHOVELER   17–20″

Note the *spoon-shaped bill,* rufous sides, and pale blue wing patches.

AMERICAN WIGEON   18–23″

Note the *white crown* and, in flight, the *white patch* on the forewing.

BLUE-WINGED TEAL   15–16″

A little duck. Note the *white facial crescent* and *pale blue* wing patches. The **Green-winged Teal** (not shown) lacks the facial crescent and blue wing patches.

WOOD DUCK   17–20″

This gaudy duck may be known by its unusual face pattern and swept-back crest. It often perches in trees.

MALLARD

female

male

AMERICAN BLACK DUCK

COMMON PINTAIL

NORTHERN SHOVELER

AMERICAN WIGEON

BLUE-WINGED TEAL

WOOD DUCK

Bay Ducks, unlike marsh ducks, dive for a living, but like the marsh ducks, most raise their families in wetlands of the northern interior. They migrate to coastal bays in winter.

LESSER SCAUP ("BLUEBILL") 15–18"

Scaups are "black at both ends and white in the middle." In winter they raft in large flocks in bays along the coast and on some lakes, retreating to northern latitudes to breed. The Lesser Scaup has a *purplish gloss* on its black head. The **Greater Scaup** (not shown) is whiter with a *greenish* gloss on its head. Both show conspicuous *white* wing stripes in flight.

CANVASBACK 20–24"

This familiar *chestnut-headed* bay duck is known from the Redhead by its whiter body and *longer, more sloping* profile. The Redhead has a more rounded head.

REDHEAD 18–23"

Grayer than the Scaup, this bay duck may be known by its *rusty red head* and *gray* wing stripe (in flight).

GOLDENEYE 20"

The *round white face spot* on the black (green-glossed) head identifies this white-bodied duck. It breeds on wooded northern lakes and winters southward wherever ice-free waters permit.

BUFFLEHEAD 13–15"

A small, chubby whitish duck with a *puffy white head patch* (see Hooded Merganser, p. 29). On Canadian lakes in summer; mainly on coastal bays in winter.

RUDDY DUCK 15–16"

In breeding season on their prairie ponds, male Ruddy Ducks are a *deep rusty red* with *white cheeks*, and swim about cocking their spiky tails. In winter, when these ducks raft in coastal bays, the rust color is replaced by gray, somewhat the color of the females.

LESSER SCAUP

CANVASBACK

REDHEAD

GOLDENEYE

BUFFLEHEAD

RUDDY DUCK

SCOTERS are heavy, blackish sea ducks, seen swimming in large rafts offshore or flying over the waves in stringy lines. The males, shown here, are black; females are dusky brown. Scoters nest in the far north and migrate and winter along both coasts.

SURF SCOTER 19″

Know this one by its *white head patches*, which have given the bird the nickname "Skunk-head." Along the West Coast it can often be seen close at hand, in harbors and around piers.

BLACK SCOTER 18½″

All black except for the *orange-yellow knob* on the bill ("Butter-nose").

WHITE-WINGED SCOTER 21″

Black with a *white patch* on each wing. These patches may be concealed when the birds are swimming.

SURF SCOTER

BLACK SCOTER

WHITE-WINGED SCOTER

MERGANSERS, or "Sawbills," are slim diving ducks with slender sawtoothed bills, unlike the broad flattened bills of other ducks. The females have crested rusty heads.

RED-BREASTED MERGANSER   20–26"
The Red-breast is more addicted to salt water than the other mergansers. The male is identified by its *wispy crest, white neck,* and reddish chest.

HOODED MERGANSER   16–19"
The *fanlike white hood* with its black border is distinctive. The Hooded Merganser prefers fresh water. Do not mistake it for the Bufflehead (p. 27), which has a white body.

COMMON MERGANSER   22–27"
The *long whitish body,* black back, *green-black head,* and thin red bill identify this resident of Canadian lakes, rivers, and streams. It winters in the U.S. and southern Canada, wherever ice-free waters permit.

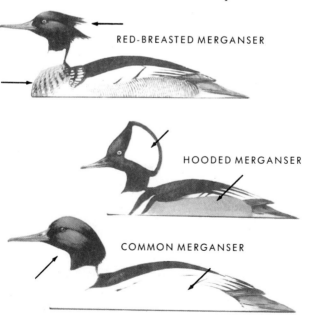

RED-BREASTED MERGANSER

HOODED MERGANSER

COMMON MERGANSER

DUCKLIKE SWIMMERS (COOTS, GALLINULES).

These birds may seem ducklike as they swim, but are really related to the rails—hen-like marsh birds that are more often heard than seen. Coots and gallinules are found widely on most continents; we have 3 in North America.

AMERICAN COOT   13–16″

Ducklike; slaty with a blackish head and neck and a hen-like *white bill*. Coots are summer residents of the freshwater ponds and marshes of the central and western parts of the continent and often gather in large flocks on the bays of both coasts in winter. They may dive for vegetarian food, or they may forage on shore.

MOORHEN (COMMON GALLINULE)   13″

Unlike the Coot, with which it sometimes associates in the marshes, the Moorhen or "Common Gallinule" (as it was formerly called) has a *bright red bill*. When walking it flirts its white undertail coverts and when swimming it pumps its head. It is widespread but furtive in the reedy ponds and marshes of the eastern half of the continent, but more local in the West. A more colorful species, the **Purple Gallinule** (not shown), is endowed with a deep purple body and *pale blue* forehead shield. This bird is a resident of the southeastern swamps. One of the best places to look for it is Everglades National Park.

Coots skitter

lobed foot of Coot

AMERICAN COOT

MOORHEN

AMERICAN COOT   MOORHEN

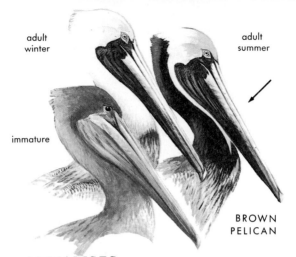

adult
winter

adult
summer

immature

BROWN
PELICAN

AERIALISTS
BROWN PELICAN 50"

Pelicans are huge robust waterbirds with *long flat bills* and great *throat pouches*. Of the 6 species in the world, 2 live in North America. This species, the Brown Pelican, is confined to salt water along our southern coasts, where it is well known to every tourist. These ponderous birds fly in orderly lines low over the water and often perch on the pilings of fishing wharfs. The **White Pelican** (not shown) ranges widely across the western half of the continent.

NORTHERN GANNET

GANNETS AND BOOBIES are large seabirds with big tapering bills and pointed tails. Gannets live in cold seas, boobies in tropical seas. Of the 9 species in the world, only one (the Northern Gannet) breeds in North America, but 4 boobies are occasional visitors to our southern waters.

NORTHERN GANNET   38″

This goose-sized seabird lives in the North Atlantic, migrating to Florida and the Gulf from its half-dozen large breeding colonies in the Gulf of St. Lawrence and Newfoundland. Larger than any gull, it can be identified by its *"pointed* at both ends" look. Adults are snow white with *jet-black wing primaries.* Young birds are dusky, darker than any young gull. Gannets scale over the waves well offshore and may be seen plunging beakfirst into the water in an ungull-like manner.

BLACK-BACKED GULL

GULLS. See family account on next page.

BLACK-BACKED GULL   28–31″

This large gull is common along the North Atlantic Coast. It may be picked out in flocks of lesser gulls by its *black back.* In the West its dark-backed counterpart, the **Western Gull** (not shown), hovers around the fish piers looking for a handout.

GULLS, long-winged and graceful in flight, are also competent swimmers. They will eat almost anything, including garbage, and thus will survive. Young gulls, running the gamut of browns, grays, and white, can be difficult to identify. There are 45 species of gulls in the world; 20 nest in North America, another 4 are visitors.

HERRING GULL   23–26″

This gray-mantled gull, almost continent-wide in its occurrence, is perhaps the best-known gull. Its black wingtips, yellow bill with a red spot, and *pale pinkish feet* are distinctive. Young birds are dusky brown.

RING-BILLED GULL   19″

This somewhat smaller edition of the Herring Gull is distinguished by the *black ring* on its bill and by its *dull yellowish* rather than pale pinkish feet. It has undergone an explosive increase in recent years, partly because it adapts readily not only to refuse dumps but also to beach developments such as restaurants and hamburger stands where there is always a handout for enterprising gulls. The Ring-bill is found seasonally along both coasts and on inland waters.

LAUGHING GULL   16–17″

This small, *hooded* gull is abundant along the Atlantic Coast, where it adds life and aerial movement to the bays and beaches. It is the only gull that breeds in the southern states. Toward winter, after it has lost its "hood," it is joined by Herring and Ring-billed Gulls from the North. At that season, know it from the others by its *dark feet*, dark bill, and smoky gray back. On the prairies a similar hooded species, **Franklin's Gull** (not shown), is found during the summer.

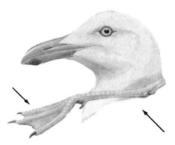

HERRING
GULL

adult

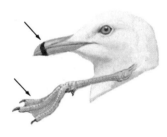

RING-BILLED
GULL

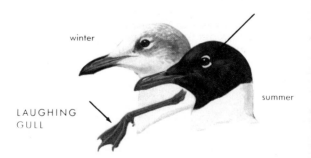

winter

summer

LAUGHING
GULL

TERNS, often nicknamed "sea-swallows," are closely related to the gulls, from which most species can be told by their *deeply forked tails, black caps,* and *sharply pointed bills.* There are 42 species in the world; 14 in North America. **Skimmers** belong to a related family. There are 3 species in the world; 1 in North America.

COMMON TERN   13–16″

When fishing, terns hover on beating wings in one spot over the water until their prey is located, then dive beakfirst. The Common Tern, shown here, is the best-known tern in the East, **Forster's Tern** in the West and South, the **Arctic Tern** in northern oceans.

LITTLE TERN   9″

Half the size of the Common Tern, the Little Tern is identified by its *yellow bill* and *wedge of white* on its forehead. Loose colonies of these "little guys" can be found on sandy beaches from New England south along the entire Atlantic and Gulf coasts, as well as in California and on sandbars of the lower Mississippi and Missouri.

BLACK TERN   14″

In summer this graceful, *black-bodied* tern hovers over inland marshes where it nests, and before summer is over it migrates through the interior and along the coasts to its tropical winter home. In fall and winter the black body plumage is replaced by white.

BLACK SKIMMER   16–20″

Along the Atlantic Coast the Skimmer plows the shallow waters with its strange bill and rests in flocks of its kind on sandy beaches. Its black-and-white pattern and curious red-and-black *bill* identify it.

winter

summer

COMMON TERN

winter

summer

BLACK
TERN

LITTLE TERN

BLACK SKIMMER

LONG-LEGGED WADERS

HERONS, EGRETS, BITTERNS. These
medium to large wading birds are endowed
with long necks, long legs, and spearlike
bills. There are 59 species in the world; 12
in North America.

GREAT BLUE HERON   42–52″
A lean, gray wading bird, 4 feet tall. It is
sometimes miscalled a crane, but unlike the
Sandhill Crane, it does not fly with its
neck extended. Great Blues breed locally in
colonies from southern Canada to Mexico,
withdrawing in winter from the colder
regions.

LITTLE BLUE HERON   24″
This slender heron, half the size of the
Great Blue Heron, is known by its *all-dark*
look (see Green-backed Heron, p. 40).
Although confined to coastal islands,
swampy southern lowlands, and the lower
Mississippi Valley when nesting, it is an
irregular wanderer to the northern states.

SNOWY EGRET   20–27″
A white heron with *"golden slippers."*
Whereas the Snowy has a *black bill* and
yellow feet, the Great Egret, twice as large,
has a yellow bill and black feet. Although
the Snowy now breeds as far north as New
England, it is basically a bird of the South.

GREAT EGRET   38″
This stately egret, nearly the size of the
Great Blue Heron, can be separated from
the smaller Snowy Egret by its *long yellow
bill*. Although mainly a summer resident
of the South, some Great Egrets now breed
colonially as far north as southern New
England and the upper Mississippi Valley.
Caution: The **Cattle Egret** (not shown) also
has a *yellow bill*, but is much smaller,
with yellow, greenish, or dusky (not
blackish) legs.

GREAT
BLUE
HERON

LITTLE
BLUE
HERON

GREAT EGRET

SNOWY EGRET

BLACK-CROWNED NIGHT HERON 23–28″

This stocky, *black-backed, pale-bodied* heron with a *black crown* is most active at dusk, when its loud *quok* proclaims its presence as it flies from its daytime roost to its nocturnal feeding ground on the marsh. Young birds are brown, suggesting the American Bittern.

GREEN-BACKED HERON 16–22″

In flight, with its neck pulled in, this small dark heron of the ponds and streams looks almost crow-like, but it flies with more bowed wingbeats. Its voice is a loud *skyow* or *skewk*. Widespread across the U.S. in summer, it enters Canada only in southeastern Ontario. It winters in Florida and along the Gulf Coast.

AMERICAN BITTERN 23″

This stocky brown heron of the marshlands looks superficially like a young Night Heron but is warmer in tone and, in flight, shows *blackish* in the outer wing. Note also the *black neck stripe.* In spring the Bittern delivers a strange "pumping" song, a slow, deep *oong-ká-choonk, oong-ká-choonk, oong-ká-choonk,* etc. It frequents the cattail marshes seasonally from central Canada to the Gulf states, withdrawing to the southern coastal marshes in winter. Poised motionless with its *bill upturned,* this bird can be almost impossible to spot among the reeds and tall marsh grasses.

Herons fly with necks pulled in, legs trailing.

adult

BLACK-CROWNED
NIGHT
HERON

immature

GREEN-
BACKED
HERON

AMERICAN
BITTERN

SMALLER WADERS

PLOVERS are small to medium-sized waders
of the beaches, marshes, and mudflats.
They are more compactly built than most
sandpipers, with shorter, pigeonlike
bills and larger eyes. There are 63 species
in the world; 10 breed in North America, 4
more are rare strays.

KILLDEER   9–11″
Widespread in farming country, the killdeer
can be identified by the *2 black bands*
across its breast and its loud, insistent
protests—*kill-deaah* or *dee-ee.* Killdeers
winter in the relatively snow-free South and
are harbingers of spring in the North.

SEMIPALMATED PLOVER   6½–7½″
Resembling a pint-sized Killdeer, this small
plover has a *single* breastband. Its legs are
orange or yellow. Whereas this little
shorebird is the color of *wet* sand, the
Piping Plover (not shown), is the color of
dry sand. The note of the "Semi" is a
plaintive upward-slurred whistle, *too-li.*
After nesting in the Arctic it migrates
across country to the coast.

BLACK-BELLIED PLOVER   10½–13½″
This large shorebird with its *black belly* is
handsome in breeding plumage, quite in
contrast to its dull coloration in winter. Its
call is a plaintive slurred whistle, *whee-
er-ee.* The Black-belly breeds in the high
Arctic but spends most of the year along
the coasts of the lower 48 states.

RUDDY TURNSTONE   8–10″
The *harlequin pattern* marks this robust,
orange-legged wader, which looks rather
like a plover but is an unusual kind of
sandpiper (see p. 44). Adults in winter
become quite brown, but retain enough of
their pattern to be recognized. Turnstones
breed in the Arctic but are found along
both coasts during migration and winter.

KILLDEER

SEMIPALMATED
PLOVER

BLACK-BELLIED
PLOVER

RUDDY
TURNSTONE

SANDPIPERS. Typical sandpipers have more slender bills than plovers. The Ruddy Turnstone, shown on the previous page, was formerly assigned to the plover family but is now regarded as more closely related to the sandpipers. There are 80 species of sandpipers and their relatives in the world; 41 nest in North America, another 20 are occasional strays from Europe or Asia.

WILLET   14–17″

This large shorebird is plain at rest and spectacular in flight, when it is immediately recognized by its flashing *black-and-white* wing pattern. The Willet is shaped somewhat like a Yellowlegs, but is larger and stockier, with *blue-gray legs.* It is commonest in marshes along the Atlantic and Gulf coasts and the shores of prairie lakes in the West, where its musical *pill-will-willet* is a familiar sound. In winter it flocks along the coasts of the southern states.

GREATER YELLOWLEGS   14″

Yellowlegs (2 species) are slim gray shorebirds distinguished by their *bright yellow legs.* In flight they are dark-winged with a whitish rump and tail. The Greater Yellowlegs has a clear 3-note whistle— *whew-whew-whew,* or *dear! dear! dear!*

LESSER YELLOWLEGS   10–11″

This slim wader of the muddy shores and marshes can be separated from its larger relative by its size and *shorter, slimmer bill.* The best clue is its call, a 1- or 2-note *yew* or *yu-yu,* less forceful than the 3-note whistle of the Greater Yellowlegs. Both yellowlegs nest in lightly timbered bogs and muskegs in Canada and Alaska, migrating throughout the continent to southern coasts.

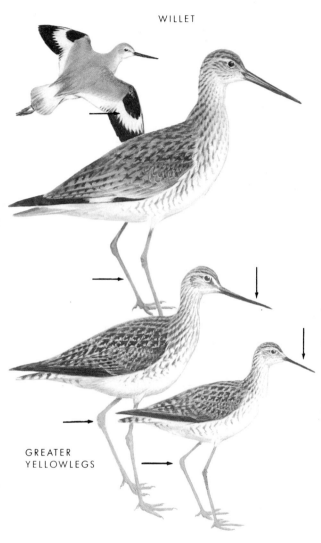

WILLET

GREATER
YELLOWLEGS

LESSER YELLOWLEGS

SPOTTED SANDPIPER 7½"

The "Spotty" inhabits the shores of lakes, ponds, and streams across Canada and much of the U.S. Identify it by the *thrushlike spots* and incessant *tail-bobbing.* In winter on southern beaches it lacks the spots. Its call is a clear *peet-weet.*

SANDERLING 7–8"

Flocks of these plump little shorebirds of the outer beaches chase the retreating waves, like clockwork toys. In flight they show bold *white wing stripes.* When breeding in the Arctic, Sanderlings are *bright rusty,* but in fall or winter they are the *palest* of the sandpipers—gray above, with *black shoulders.*

DUNLIN 8–9"

In springtime the Dunlin is smartly dressed, with a *rusty red back* and a *black patch* across its belly. In fall and winter it is more soberly attired; then note the *droop* toward the tip of its bill. Dunlins breed in the high Arctic and migrate to the coast, where they tarry on the tidal flats.

LEAST SANDPIPER 5–6½"

A sparrow-sized sandpiper of mudflats and marsh edges. The *browner* color, *thinner bill,* and smaller size separate it from the similar **Semipalmated Sandpiper** (not shown), which prefers more open beaches. Breeding in Canada and Alaska, the Least Sandpiper retreats to the Gulf states in winter.

AMERICAN WOODCOCK 11"

Near the size of a Bobwhite, the Woodcock is distinguished by its *very long bill* and pop eyes. When it flies it whistles with its rounded wings. The Woodcock is widespread in wet thickets and brushy swamps east of the plains, from Canada to the Gulf states.

winter

summer

SPOTTED SANDPIPER

SANDERLING

summer

winter

DUNLIN

LEAST
SANDPIPER

AMERICAN
WOODCOCK

FOWL-LIKE BIRDS

Grouse, often misnamed "partridges," are chicken-like, larger than quail and lacking the long tails of pheasants. There are 18 species in the world, 10 in North America. **Pheasants, quail,** and **true partridges** are often lumped in one large family of 165 species; we have 10, including 4 introduced species.

RUFFED GROUSE   16–19″

This chicken-like bird of the woodlands with its *fan-shaped tail* flushes with a whir. Two color phases occur: gray and reddish. The latter is more frequent in the southern parts of its range, which extends across wooded Canada and south in forested country to the central Appalachians. Gray birds are typical of the Rockies; reddish birds of the Pacific states. The male's "drumming" suggests a distant motor starting up. The muffled thumping of the wings starts slowly, then accelerates into a whir.

COMMON BOBWHITE   8½–10½″

This small, rotund game bird, a favorite of sportsmen, is near the size of a Meadowlark. The *"bob-white"* call is familiar in agricultural country and along hedgerows east of the Rockies, except in the northernmost states.

RING-NECKED PHEASANT
30–36″ (male), 21–25″ (female)

This gamecock-like bird with its sweeping tail is an introduction from the Old World. Males are gaudy, females are brown; unlike the Ruffed Grouse, Pheasants have *long pointed tails.* Pheasants thrive in farm country across the northern states and the adjacent parts of Canada, capitalizing on waste grain left by the reapers.

gray phase

red phase

RUFFED GROUSE

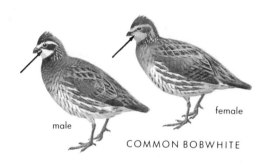

male

female

COMMON BOBWHITE

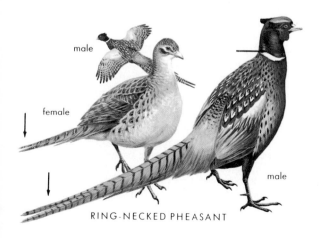

male

female

male

RING-NECKED PHEASANT

BIRDS OF PREY

 FALCONS are streamlined birds of prey with long pointed wings and longish tails. There are 52 species in the world; we have 6, plus 2 accidentals from the Old World.

PEREGRINE FALCON   15–20″

The Peregrine can be recognized from other falcons by its size (near that of a Crow) and its heavy *blackish "sideburns."* This favorite of falconers was nearly eliminated from much of North America when DDT was used, but is now being restored by artificial propagation and restocking. During its travels the Peregrine favors open country, especially along the coast.

AMERICAN KESTREL   9–12″

Shaped like a large swallow, but the size of a jay, the Kestrel is known by its sharply pointed wings and longish tail. Males have blue-gray wings and a *rusty back and tail.* Females lack the bluish on the wings. This is a common bird along country roads, where it often uses poles or wires as lookout perches. When scouting for mice and grasshoppers the Kestrel (formerly called "Sparrow Hawk") often hovers in one spot in the air, searching the ground below for prey. Its voice is a high, excited *killy killy killy.*

HARRIERS are slim hawks with slim wings and long tails.

NORTHERN HARRIER   17½–24″

This large, slender bird of prey, often seen gliding low over the meadows and marshes with its wings held in a shallow V, can be identified by its conspicuous *white rump patch.* Males are pale gray above; females are brown. The Harrier breeds across Canada and the northern half of the U.S. It winters from the northern states to northern South America.

PEREGRINE FALCON

AMERICAN KESTREL

female

male

male

immature

NORTHERN HARRIER

male

female

BUTEOS (BUZZARD HAWKS) are large, thick-set birds of prey with broad wings and wide, rounded tails. They habitually soar aloft in wide circles. There are 10 species in North America.

RED-TAILED HAWK   19–25″

When this large, broad-winged, round-tailed hawk banks in its soaring, the *rufous* of the upperside of the tail can be seen. (Young birds do not have red tails.) When this hawk is perched, its *white breast* contrasts with its dark head and the band of dark streaks across its belly. Red-tails live from timberline in Canada and Alaska to Panama, but retreat in winter from northern forests. The voice is a squealing *keeer-r-r.*

RED-SHOULDERED HAWK   17–24″

This buteo, with its ample tail and broad wings, differs from the Red-tail by its *banded tail* and *reddish underparts.* The *reddish "shoulders"* are also a good mark. Its cry is a slurred, 2-syllabled scream, *kee-yer.* Although mainly a hawk of the East and Midwest, it is also found locally in California.

ACCIPITERS (BIRD HAWKS) are long-tailed woodland hawks with short, rounded wings and longish tails. Typical flight is several quick beats and a glide. There are 3 species in North America.

SHARP-SHINNED HAWK   10–14″

The *short, rounded wings* make it possible for the "Sharpie" to pursue its prey through the trees. It is found continentwide at one season or another, and is the commonest species observed at the hawk lookouts such as Hawk Mountain, Pennsylvania and Cape May, New Jersey. **Cooper's Hawk** (not shown) is a larger version.

RED-
TAILED
HAWK

immature

adult

RED-SHOULDERED
HAWK

adult

SHARP-SHINNED
HAWK

immature

adult

AMERICAN VULTURES. Often incorrectly called "buzzards," these eagle-like birds can often be seen soaring aloft in wide circles. Their rather small heads are naked of feathers. There are 7 species in the world; 3 live in North America.

TURKEY VULTURE 26–32"

With a wingspan of 6 feet—nearly that of an eagle—this scavenger soars over the landscape looking for carrion. Its small head is covered with wrinkled *red skin*. As it glides along, rocking or tilting unsteadily, it holds its wings in a shallow V or dihedral. Ranging from southern Canada to the tropics, it retreats in winter from the snow-covered parts of the continent.

The **Black Vulture** (not shown) is more southern. It has a *black* head and conspicuous white patches toward the wingtips.

TURKEY VULTURE

OSPREY

OSPREY 21–24″

The Osprey or "Fish Hawk" is a familiar sight along some sections of the Atlantic Coast, where its big bulky nests can be seen even from train windows, or from well-traveled roads. It is also found locally around the Great Lakes, across Canada, and locally in the western states. This big fish-eating raptor hovers over the water on laboriously beating wings, then dives feet-first for its prey. In flight, the Osprey shows a *dark patch* on the underside of its wings. It also has a *broad black cheek patch*.

Ospreys build bulky nests in trees and may use platforms put up on poles for their use.

OWLS —most of them—are nocturnal birds with large puffy heads, flat faces, and forward-facing eyes. Like other birds of prey, they have hooked bills and hooked claws, but the feet are usually feathered and the outer toe is reversible. 134 species in the world; 18 in North America.

EASTERN SCREECH OWL 7–10″

Screech Owls of one sort or another are found from Canada to the tropics. Those of western North America have been separated from those of the East as a different species. Eastern Screech Owls come in two color phases, red-brown and gray. The **Western Screech Owl** is usually gray, but some birds of the Great Basin or the Pacific Northwest may be brown. The voice of the eastern bird is a mournful whinny, descending in pitch. Sometimes it utters a series of notes or whistles on one pitch, rather like the call of the western bird.

BARN OWL 16–20″

This owl, an inhabitant of barns and belfries, is the "Monkey-faced Owl," recognized by its *dark eyes* set in a *white heart-shaped face.* At night its whitish or pale cinnamon underparts and silent mothlike flight give it a ghostly look as it cruises the meadows for mice. It does not hoot like the larger owls but has a shrill, rasping hiss or snore: *kschh* or *shiiish.*

GREAT HORNED OWL 18–25″

This *large* owl of cat-like mien is distinguished by its size, prominent *ear tufts, white bib,* and heavily *barred* breast. The hooting of the male—*hoo-hoo-oo, hoo hoo*—is often answered by the lower-pitched hooting of the larger female. Great Horned Owls are found from tree limit in Canada to the southern tip of South America.

EASTERN
SCREECH
OWL

gray phase

red-brown phase

BARN OWL

GREAT
HORNED
OWL

NONPASSERINE (NON-PERCHING) LAND BIRDS

PIGEONS AND DOVES. These fast-flying birds with cooing voices come in two basic types, exemplified here by the Rock Dove, with its fanlike tail, and by the more slender Mourning Dove. There are 289 species in the world; 11 in North America (plus 6 strays).

ROCK DOVE (DOMESTIC PIGEON)   13″
The typical Rock Dove or park pigeon, familiar to city dwellers, shows a *white rump* and *2 black wing bars* when it flies, but the domestic stock may come in many color variants.

MOURNING DOVE   12″
This, the widespread wild dove, is smaller and slimmer than the Rock Dove. Its *pointed tail* shows large white spots as the bird flies away. It often comes to feeders. Its voice is a mournful *coah, cooo, cooo, coo.*

ROCK DOVE

MOURNING DOVE

CUCKOOS (AND ALLIES). Slender, long-tailed birds whose feet grasp branches with 2 toes forward, 2 aft. There are 128 species in the world; 6 live in North America.

YELLOW-BILLED CUCKOO　　　　　11–13"

Long and slim, our two common cuckoos can be recognized by their brown backs and whitish underparts. They are shy and secretive. This species has a *yellow bill; large, white tail spots;* and *rufous* primaries in the wings. Its call is a rapid, throaty *ka-ka-ka-ka-ka-ka-kow-kow-kowlp-kowlp-kowlp-kowlp.* Widespread across the U.S. in summer, it winters in South America. The **Black-billed Cuckoo** (not shown) is similar but the bill is *black.* There is no rufous in the wing, and the tail spots are small. The song is a rhythmic *cucucu, cucucu, cucucu,* etc. Both cuckoos may sing at night. The Black-bill, a bit more northern, breeds across southern Canada and central and northeastern U.S. It winters in South America.

YELLOW-BILLED
CUCKOO

HUMMINGBIRDS. These, the smallest birds, are feathered jewels—iridescent birds with needle-like bills for sipping nectar. There are about 320 species in this New World family, 15 of which reach the U.S. (plus 6 accidentals). Only one, the Ruby-throat, is normally found in the East.

RUBY-THROATED HUMMINGBIRD   3–3¾"
The male Ruby-throat has a *fiery red throat* and a green back. The female lacks the red. Do not mistake sphinx moths for "baby hummingbirds!" The Ruby-throat, ranging from southern Canada to the Gulf, and from the Atlantic to the Great Plains, is our only eastern hummer. It winters in the tropics, but a few may linger in southern Florida and southern Texas.

RUFOUS HUMMINGBIRD   3½"
No other male hummer has a *rufous back.* This species replaces the Ruby-throat in much of the West, migrating through the Pacific states to its breeding grounds in northwestern North America and returning to Mexico by way of the Rockies.

RUBY-THROATED HUMMINGBIRD

male

female

male

male

RUFOUS HUMMINGBIRD

female

COMMON NIGHTHAWK

WHIP-POOR-WILL

GOATSUCKERS are well-camouflaged
nocturnal birds with large eyes, tiny bills,
large bristled gapes, and very short legs.
There are 67 species in the world, 8 in
North America.

WHIP-POOR-WILL 9½″

The Whip-poor-will is a voice in the dark, a
tirelessly repeated *whip' poor-weel'*,
whip' poor-weel , etc. Rarely is the singer
glimpsed unless flushed like a large brown
moth from the leafy forest floor, showing
its rounded wings and *white or buff
tail patches.* It is found during summer in
the northeastern and central parts of the
continent. In winter it joins its southern
relative, the **Chuck-will's-widow** (not
shown), in the Gulf states.

COMMON NIGHTHAWK 9½″

The Nighthawk (not a hawk) is usually seen
cruising for insects high in the air, where
it flies with easy strokes, "changing gear" to
quicker erratic strokes. On warm summer
evenings the male in aerial display makes
steep dives. Note the *broad white bar
across* the pointed wings. Its voice is a
nasal *peent* or *pee-ik.* Widespread in
summer, it winters in South America.

KINGFISHERS are solitary birds with large heads, daggerlike bills, and tiny feet. Most (but not all) species eat fish. There are 87 species in the world; only 3 occur in North America.

BELTED KINGFISHER 13"

The Belted Kingfisher gives voice to a loud dry rattle as it flies over the water. This big-headed, big-billed bird is larger than a Robin and is blue-gray above with a *ragged bushy crest* and a *broad gray breastband*. The female has an additional *rusty breastband*. When fishing, this expert nimrod hovers over the water so as to spot its prey, then dives beakfirst, not always with success. Found widely across much of North America, it may winter as far north as ice-free fresh water permits.

female

BELTED
KINGFISHER

WOODPECKERS. These chisel-billed, wood-boring birds have strong feet (usually 2 toes in front, 2 in rear) and stiff, spiny tails that act as props when the bird climbs. Most males have some red on the head. There are 210 species in the world; 21 in North America.

DOWNY WOODPECKER   6½″

Nearly everyone who puts out suet knows the Downy, the little spotted woodpecker with a *white back* and a relatively *small bill*. It occupies a vast range from coast to coast and from near the treeline in Canada and Alaska to the Gulf Coast. Its call is a rapid whinny, descending in pitch; its note is a flat *pik*.

HAIRY WOODPECKER   9½″

Half again as large as the Downy Woodpecker, the Hairy can be recognized by its *much larger bill*. Males of both species have a small red patch on the back of the head which the females lack. The ranges of these two woodpeckers are relatively similar. The call of the Hairy is a loud, kingfisher-like rattle; its note is a sharp *peek!*

HAIRY
WOODPECKER

male

male

DOWNY
WOODPECKER

COMMON FLICKER 12–14″

When this large, brown-backed woodpecker flies, it shows a conspicuous *white rump* and a *black crescent* across its chest. We often see this bird on the ground, where it eats ants. Flickers are found almost continentwide except in treeless country. They come in two basic forms: Yellow-shafted (East), with yellow linings under the wings and tail; and Red-shafted (West), with salmon-red linings. Where the two forms meet on the Great Plains they hybridize.

RED-HEADED WOODPECKER 8½″–9½″

The *entire head is red* in this unmistakable woodpecker of the shade trees, orchards, and groves of farms and villages. Its large white wing patches make the lower back look white. The note, coming from the oaks, is a loud *queer* or *queeah*. Redheads range from southern Canada to the Gulf of Mexico and west to the Great Plains, but are absent from New England.

RED-BELLIED WOODPECKER 9–10½″

In suitable woodlands and groves from the Great Lakes and Connecticut south to the Gulf states and west to the edge of the prairies, the Red-bellied Woodpecker is a year-round resident. The male, with his *zebra back* and *scarlet crown*, is unmistakable. In the female the red is confined to the nape. The notes—*churr*, or *chiv, chiv*—often betray this bird's presence among the trees.

ACORN WOODPECKER 8–9½″

The *clownlike face* and *black back* identify this woodpecker of the oak groves and wooded canyons of the West. It is resident from Oregon and the southwestern states to Panama.

male

red-shafted
form

male

COMMON
FLICKER

yellow-shafted
form

adult

male

RED-BELLIED
WOODPECKER

RED-HEADED
WOODPECKER

male

ACORN
WOODPECKER

PASSERINE (PERCHING) BIRDS

TYRANT FLYCATCHERS. Most flycatchers perch upright on exposed branches, from which they dart forth to snap up flying insects. 365 species, all in the New World; most live in the tropics, but there are 33 in North America, plus at least 5 accidentals.

EASTERN KINGBIRD   8″
Know the Kingbird by the wide *white band* at the tip of its tail. This bicolored bird often harasses crows (a habit shared with the Red-winged Blackbird). It is a familiar summer resident of farming country, favoring orchards and fencerows, where it sits conspicuously on roadside wires. Its voice is a rapid sputter of high bickering notes. Widespread from central Canada to the Gulf states, it winters in South America.

WESTERN KINGBIRD   8″
Found mainly west of the Mississippi Valley, this large flycatcher has habits similar to those of the Eastern Kingbird. Like that bird, it favors roadsides in farming country, where it sits conspicuously on wires. The grayish chest, *yellowish belly*, and *black tail* identify it. Although this bird of the West winters mainly in Central America, a very few Western Kingbirds wander eastward in autumn.

GREAT CRESTED FLYCATCHER   8–9″
This kingbird-sized flycatcher can be identified by its bushy crest, *yellowish belly*, and *cinnamon-toned* wings and tail. Its cry, coming from the woodlands, is a loud whistled *wheep!* Widespread across southern Canada and the U.S. east of the Rockies, it winters in the tropics. Two similar species live in the Southwest.

EASTERN
KINGBIRD

WESTERN
KINGBIRD

GREAT CRESTED
FLYCATCHER

EASTERN PHOEBE 6½–7″

From the Great Plains to the Atlantic, the modest Phoebe frequents farms and streamsides, building its nest under the eaves of a shed or under a bridge. The habit of *bobbing its tail* is a giveaway, as is its call, a well-enunciated *fee-be.* Hardier than its relatives, it spends the winter in the southern tier of states and returns northward relatively early, often to be inconvenienced in its quest for insects by a late snowfall.

EASTERN PEWEE 6–6½″

The Pewee is very much like the Phoebe, except that it has *2 well-defined wing bars.* Basically a woodland bird, its call is a plaintive slurred whistle, *pee-a-wee.* Its nest, unlike that of the Phoebe, straddles a horizontal branch in a tree. Widespread east of the plains, from southern Canada to the Gulf, the Pewee retires to the tropics for the winter. The **Western Pewee** (not shown) is similar, but sings a nasal *pee-yee.*

LEAST FLYCATCHER 5¼″

The Least Flycatcher, a bird of farm groves and orchards, is sometimes called "Chebek," because of its emphatic *che-bek!* call. It is one of 8 or 9 similar small flycatchers with *light eye-rings* and *wing bars* in the U.S. To straighten out these birds (sometimes a very difficult task), you must study *A Field Guide to the Birds* or *A Field Guide to Western Birds.*

SAY'S PHOEBE 7–8″

This rather large phoebe of the West is identified by the *rusty wash* on its underparts, which gives it the look of a small Robin. Its voice is a plaintive *pee-ur* or *pee-ee.*

EASTERN
PHOEBE

LEAST
FLYCATCHER

EASTERN
PEWEE

SAY'S
PHOEBE

SWALLOWS. Streamlined form and graceful flight characterize these sparrow-sized birds. Swallows often sit in rows on wires, where each can be recognized from below by the cut of its jib and its markings. Of the 75 species of swallows in the world, 8 normally live in the United States and Canada; 5 others occur as strays.

BARN SWALLOW   6–7¾″

The *forked tail* with its streaming outer feathers distinguishes the Barn Swallow from other members of its family. Its nest is an open-topped mud cup on a beam or support, usually *inside* the barn. The Cliff Swallow (below) places its mud jugs under the eaves *outside* the barn. The Barn Swallow is a widespread summer resident in North America except in the Southeast, where its range is slowly expanding. Inasmuch as it depends on flying insects for its food, the Barn Swallow, like most of the other birds in its family, retreats to the tropics for the winter.

CLIFF SWALLOW   5–6″

Note the *pale buffy rump.* In lieu of cliffs, which it uses in some regions, this colonial species often builds its jug-like mud nests under the eaves of barns. Nesting widely in North America except in the Southeast, it retreats in winter to South America.

PURPLE MARTIN   7½–8½″

Males of this, our largest swallow, are uniformly *blue-black above and below;* females have grayish underparts. These days Martins nest mostly in bird houses with a number of compartments. Martins winter in South America and return very early to the Gulf states, whence some proceed as far north as Canada.

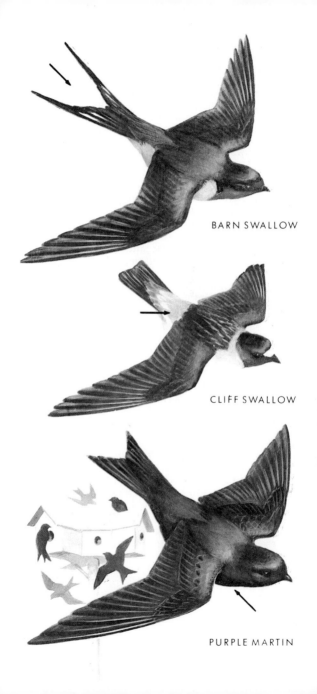

BARN SWALLOW

CLIFF SWALLOW

PURPLE MARTIN

MORE SWALLOWS AND A SWIFT

TREE SWALLOW   5–6"

This trim swallow, steely blue-green-black above and *clear white below*, is readily enticed to nest in the right kind of bird box. Widespread in the northern U.S. and Canada, it winters along southern coasts where it may feed on bayberries and similar fare when there are no insects.

BANK SWALLOW   4½–5½"

This small *brown-backed* swallow can be identified by the *dark band* across its breast. Although it breeds widely in North America its colonies are very local, found only where there are high river banks or sand quarries with soil of the right consistency for digging nest burrows. The other brown backed species, the **Rough-winged Swallow** (not shown), has a *dusky throat* and lacks the breastband.

SWIFTS.

Although swifts look rather like swallows, they are unlike them structurally, with flat skulls and all 4 toes pointing forward. In flight swifts sail with wings stiffly *bowed* between rapid spurts. There are 79 species in the world; 4 live in North America, 5 others have been recorded as strays.

CHIMNEY SWIFT   5–5½"

"Like a cigar with wings." This blackish, swallowlike bird with *no apparent tail* has long, slightly *bowed,* stiff wings that seem to twinkle as the bird flies. The rapid chippering notes are distinctive. It nests and roosts in chimneys, leaving in autumn for its winter home in Amazonian Peru. This is the only swift in the East; in the West it is replaced by the smaller **Vaux Swift** (not shown).

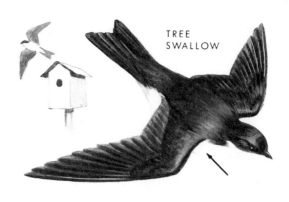

TREE
SWALLOW

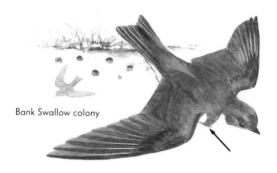

Bank Swallow colony

BANK SWALLOW

CHIMNEY SWIFT

CROWS, MAGPIES, JAYS. Crows and ravens are large and black; magpies are black and white with long tails; jays are colorful. There are 100 species in the world, 14 in North America.

AMERICAN CROW   **17–21″**

Big and black, the Crow needs no description. Its familiar *caw* or *cah* is easily imitated and its straggling flocks are familiar as they fly to their roosts. Widespread across North America, it winters as far north as deep snows allow. Along the Atlantic and Gulf coasts the smaller **Fish Crow** (not shown) is also resident.

BLACK-BILLED MAGPIE (opposite)   17–22½″

This long-tailed bird of Old World origin entered North America long ago by way of Alaska. It is widespread in the West and ranges very rarely as far east as the Great Lakes. The black-and-white pattern and *long tail* make it unlike any other bird except the **Yellow-billed Magpie** (not shown), which is restricted to the central valleys of California.

AMERICAN CROW

BLUE JAY
11–12½″

The showy and noisy Blue Jay is known by its jaunty *crest, black necklace,* and *white spots* in its wings and tail. The typical call is a raucous *jeeah* or *jay*. It is a steady customer at feeders, favoring sunflower seeds. Widespread in eastern North America, it has recently spread sparingly to the Colorado Rockies, where it met the dark blue **Steller's Jay** (not shown), a common bird of western conifer forests. The latter has a *crest* but lacks the white spots of the Blue Jay.

SCRUB JAY
11–12″

This *crestless* jay of the oaks lives in Florida and the western states, a curious case of discontinuous distribution. Its call—a rough, rasping *kwesh, kwesh*—is quite unlike that of the Blue Jay.

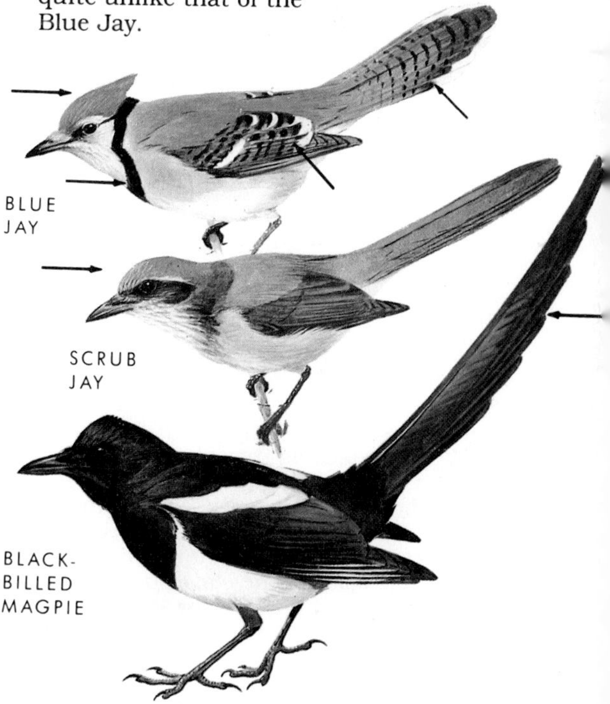

BLUE JAY

SCRUB JAY

BLACK-BILLED MAGPIE

TITMICE, ETC. These small, plump acrobats which roam the woods in little bands are easily attracted to feeders. Of the 64 species in the world, 12 are found in North America.

BLACK-CAPPED CHICKADEE 4¾–5¾″

Perhaps the most-loved bird, the Chickadee patronizes the feeder, favoring suet and sunflower seeds. Its *black cap and bib* are distinctive, as is its call, a clearly enunciated *chick-adeedeedee.* The song, a whistled *fee-bee,* is easily imitated. It lives across the southern half of Canada and the northern half of the lower 48 states.

CAROLINA CHICKADEE 4½″

Not quite a carbon copy of its northern relative, the Carolina Chickadee is basically known by its more southern range, and by its song, a 4-syllabled *fee-bee, fee-bay.* Where its range overlaps that of the Black-cap, it can be separated by its smaller size and virtual *lack of white* in the wing. Its range is from the southern edge of the Black-cap's range in New Jersey, Ohio, and Missouri, south to Florida and the Gulf.

TUFTED TITMOUSE 6″

This small, *crested* bird often joins the chickadee at the feeder. Its voice is a whistled *peter-peter-peter.* It is found from the southern Great Lakes and southern New England south to the Gulf states and west to the prairies.

BUSHTIT 4″

In the West these very plain little birds are easily overlooked as they move from bush to bush in straggling groups, talking to each other in gentle notes. Notice the small bill and the relatively *long tail.* Bushtits are widespread from Washington and Colorado south. The color of the *cheek patch* varies with location; in birds of the Rockies it is brownish.

CAROLINA
CHICKADEE

BLACK-
CAPPED
CHICKADEE

TUFTED
TITMOUSE

BUSHTIT

CREEPERS. Small, slim, stiff-tailed tree climbers with slender, *decurved bills.* There are 6 species in the world; we have 1.

BROWN CREEPER 5″

Whereas nuthatches go down trees headfirst, the Brown Creeper plays it safe, hugging the bark, starting at the base of the tree and spiraling *upward.* Creepers live in the forests across Canada, in the cooler woodlands of the northern states, and southward in the higher mountains. In winter many may travel to the Gulf states.

NUTHATCHES are small, stout tree climbers with strong, woodpeckerlike bills. Their square-cut tails are not used as braces when climbing. Unlike woodpeckers, they go down tree trunks *headfirst.* Of the 31 species in the world we have 4.

WHITE-BREASTED NUTHATCH 5–6″

This, the best-known nuthatch, is diagnosed by its *black cap* and *beady black eyes* on a *white face.* It comes readily to the suet feeder. Its note is a nasal *yank, yank.* A permanent resident of woodlands from southern Canada to the Gulf states, and across the continent from the Atlantic to the wooded West, it is absent in the treeless prairies and plains.

RED-BREASTED NUTHATCH 4½″

This small nuthatch may be known from the commoner White-breast by the broad *black line* through the eye. The call is a higher and more nasal *ank* or *enk,* sounding like a tiny tin horn. The Red-breast breeds in the cool conifer forests of Canada and the evergreen-clad ranges of the U.S., migrating some years as far south as the Gulf of Mexico.

BROWN
CREEPER

WHITE-
BREASTED
NUTHATCH

RED-
BREASTED
NUTHATCH

WRENS. Small, energetic brown birds with slim, slightly curved bills. They may cock their tails, or dip them when they sing. There are 63 species in the world; 9 in North America.

HOUSE WREN   4½–5″

This is our best-known wren, the one that shares our gardens and often nests in boxes built specifically for it. Its stuttering, gurgling song is almost incessant. Nesting across southern Canada and the northern U.S., it winters in the southern tier of states.

CAROLINA WREN   5¾″

This large wren is identified by its bright rusty color and *broad white eyebrow stripe.* Unlike the House Wren, it is sedentary, residing from the southern Great Lakes and southern New England to the Gulf, but not in the West. Its song is a 3-syllabled chant sounding like *tea-kettle, tea-kettle, tea-kettle, tea.*

BEWICK'S WREN   5¼″

This wren, which lives mainly in the Midwest and West, has a *white eyebrow,* suggesting the Carolina Wren, but note the longer tail with *white corners.* Its song suggests a Song Sparrow's, with variations. Like the House Wren, Bewick's may nest in bird boxes.

MARSH WREN   4½–5½″

Like the previous two wrens, the Marsh Wren has a conspicuous *white eyebrow stripe,* but also a *striped back.* It dwells in the marshes, where it sings its reedy, rattling song. It is widespread in summer across the northern states and southern Canada and also in some marshes along the Atlantic, Pacific, and Gulf shores.

HOUSE
WREN

CAROLINA
WREN

BEWICK'S
WREN

MARSH
WREN

KINGLETS AND GNATCATCHERS. These tiny, active "birdlets" are but small branches of the family of Old World warblers, which numbers 332 species. In North America, north of Mexico, we have only 6 species of this family, plus 4 accidental strays from Asia.

GOLDEN-CROWNED KINGLET 3½″

A tiny mite with a bright *crown patch* (*yellow* in the female, *orange* in the male). Smaller than most warblers, kinglets favor conifers but also frequent other trees in winter. The note of the Golden-crown is a high, wiry *see-see-see*, inaudible to some ears. Its song, also high and thin, drops off into a little chatter. The Golden-crown ranges in summer across forested Canada and southward in the higher mountains of the U.S.

RUBY-CROWNED KINGLET 4″

The little Ruby-crown is the kinglet with the *eye-ring*; the Golden-crown has the head stripes. The male Ruby-crown has a concealed *red crown patch* that it can expose when excited. The Ruby-crown's note is a husky *ji-dit*; the song, tentative at first, ends in a loud *ti-dadee, ti-dadee, ti-dadee.* This kinglet breeds widely across the conifer forests of Canada and Alaska, and southward through the high mountains of the western U.S. It winters widely in the southern half of the U.S.

BLUE-GRAY GNATCATCHER 4½″

The Gnatcatcher looks like a tiny Mockingbird. Smaller-bodied than a Chickadee, it is blue-gray above and whitish below, with a narrow white eye-ring and a *long black-and-white tail.* It summers in open woodlands in the eastern U.S., withdrawing in winter to the Gulf states and Mexico. It is also found in the Southwest.

GOLDEN-CROWNED
KINGLET

RUBY-CROWNED
KINGLET

BLUE-GRAY
GNATCATCHER

MOCKINGBIRDS AND THRASHERS are
often called "mimic thrushes." They
are longer-tailed than true thrushes (see p.
87) and are excellent songsters. Of the 30
species, all in the New World, we have
10, plus 1 accidental.

NORTHERN MOCKINGBIRD 9–11″

Gray and slim, longer-tailed than a Robin,
the Mockingbird flashes *large white
patches* in its wings and tail. In recent
years it has extended its range northward
as far as the Great Lakes and New England,
largely because of suburban plantings of
multiflora rose and other berry-bearing
shrubs which insure winter survival. Its
famous song is a varied series of phrases,
each repeated several times. Some Mockers
are excellent mimics.

GRAY CATBIRD 9″

The Catbird, familiar in suburban gardens,
is all-gray with a *blackish cap* and a deep
rufous patch under the tail. The catlike
mewing note is characteristic. The song is
a musical series of disjointed phrases
that are not repeated as in the Brown
Thrasher and Mockingbird. Its range—east
of the Rockies—is similar to that of the
Thrasher.

BROWN THRASHER 11½″

Longer and slimmer than a Robin, the
Brown Thrasher may be told from the
brown thrushes (p. 87) by its *longer tail*
and *striped*, rather than spotted, breast.
Related to the Catbird and Mockingbird,
this bird has a song that bears a resemblance in quality, but each phrase is
repeated in couplets, not delivered singly as
in the Catbird, nor in series as in the
Mocker. The Brown Thrasher is widespread
east of the Rockies, from southern Canada
to the Gulf states, spending the winter in
the southern parts of its range. Several
other thrashers live in the Southwest.

NORTHERN
MOCKINGBIRD

GRAY
CATBIRD

BROWN
THRASHER

LOGGERHEAD
SHRIKE

SHRIKES are songbirds with hook-tipped
bills and hawklike behavior; they prey on
mice and small birds. Widespread in the
Old World, there are 74 species; we have
only 2.

LOGGERHEAD SHRIKE 8″

This Shrike, sitting quietly on a wire or
bush top, is patterned superficially like a
Mockingbird but has a well-defined
black mask. It impales its prey—insects,
mice, and even small birds—on thorns
or barbed wire. Once found widely across
the U.S., this species has all but disappeared
from the northeastern parts of its range. A
similar species, the **Northern Shrike** (not
shown), comes down from northern
Canada and Alaska in some years to visit
the northern states.

THRUSHES (opposite page). Large-eyed,
slender-billed, strong-legged songbirds. We
show two with spotted breasts, but the
Robin and Bluebird are also thrushes.
There are 304 species in the world. 14 nest
in North America; 9 others are rare or
accidental strays.

HERMIT THRUSH

6½–7¾″

The Hermit Thrush, the divine songster of the cool forests of the North and the higher mountains, can be distinguished from the other thrushes by its *rusty tail*. Its song is clear, ethereal, and flutelike; 3 or 4 phrases at different pitches with a long pause between each phrase. At the approach of winter the Hermit deserts the forests where it nests for the southern states and is the only one of the brown thrushes likely to be found within the U.S. at that season.

WOOD THRUSH

8″

There are 5 thrushes with spotted breasts in North America. This, our best-known spotted thrush, can be heard in summer from woodlots and groves in the suburban and rural countryside of the eastern half of the U.S. It can be told from the others by the *redness* around its head and shoulders and by its larger, *rounder* spots. Its flutelike phrases are richer.

HERMIT THRUSH

WOOD THRUSH

EASTERN BLUEBIRD   7″

The gentle Eastern Bluebird is known by
its *blue back, robin-red breast* and *rusty
throat*. Females are duller. Widespread
in the East, it is replaced from the Rockies
to the Pacific by the **Western Bluebird**
(not shown), which has a *blue* throat.
Bluebirds have declined in some areas, but
can be helped by bird boxes in which they
may nest or spend the night.

MOUNTAIN BLUEBIRD   7″

This bluebird, confined to the West, is
turquoise above and *below*, shading into
whitish on the belly. The female is dull
brownish with a touch of blue in the
wings, rump, and tail.

AMERICAN ROBIN   9–11″

This big thrush of our lawns and gardens
is recognized by its *brick-red breast* and
gray back. Although the Robin has a wide
summer range, from treeline in Canada
and Alaska to the Gulf states, it is regarded
as a harbinger of spring in the northern
U.S. and Canada. Its song, a clear caroling
of short phrases, acts as an alarm clock
for many of us.

VARIED THRUSH   9–10″

In the cooler forests of the Northwest the
"Alaska Robin" sings its strange song: an
eerie, quavering whistled note, followed
after a pause by one on a lower or higher
pitch. When spotted in the wet forest this
bird has the look of a Robin with a *black or
gray band* across its rusty breast. Note
also the *orange stripe* behind the eye and
the orange wing bars. During the winter
months the Varied Thrush can be found in
woodlands throughout the Pacific states.

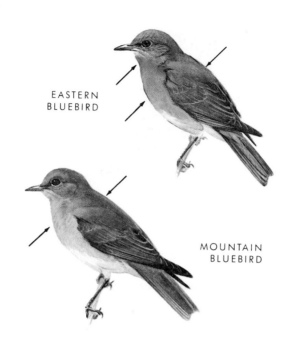

EASTERN
BLUEBIRD

MOUNTAIN
BLUEBIRD

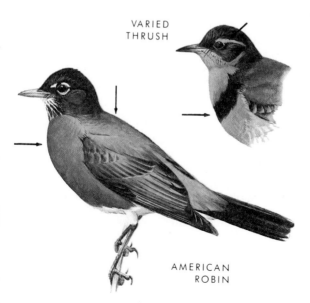

VARIED
THRUSH

AMERICAN
ROBIN

CEDAR
WAXWING

WAXWINGS are sleek, crested brown birds
with waxy red tips on their secondaries.
There are 3 species in the world; we have 2.

CEDAR WAXWING 6½–8″
The *yellow band* at the tip of the tail and
the *pointed crest* distinguish this sleek
brown bird from all but its northwestern
relative, the **Bohemian Waxwing** (not
shown), which has a *rusty patch,* not a
white one, under the tail. The note of the
Cedar Waxwing is a high, barely audible
lisp, or *zeee.* Widespread across Canada
and northern U.S., it is nomadic, traveling
as far south as Panama some winters, or
remaining as far north as southern Canada
if the food supply warrants.

VIREOS (opposite) are small, rather plain
birds, much like warblers (p. 92), but their
bills have a more curved ridge and a slight
hook. Living only in the New World, there
are 41 species; north of the Mexican border
there are 11, plus 2 accidentals.

RED-EYED VIREO 6″
This plain gleaner of insects of the woods
and shade trees is heard more often
than seen. During summer its abrupt,
robin-like phrases are repeated monoto-
nously. A look through the binocular reveals a *black-bordered white eyebrow
stripe,* a *gray cap,* and an olive back. This

vireo has no wing bars. After spending the summer east of the Rockies, from Canada to the Gulf states, it winters in the Amazon Basin.

YELLOW-THROATED VIREO 5″

The *bright yellow throat, yellow "specta-cles,"* and white wing bars identify this vireo of the woods and groves. It summers widely in the U.S. east of the Great Plains, but barely crosses the Canadian border.

SOLITARY VIREO 5–6″

Note the conspicuous *white "spectacles"* on its gray head, and the 2 white wing bars. The whistled phrases are similar to those of the Red-eye but sweeter and more deliberate. This vireo prefers woodlands with some conifers; thus it finds the cooler forests of Canada, the Appalachians, and the western mountains to its liking.

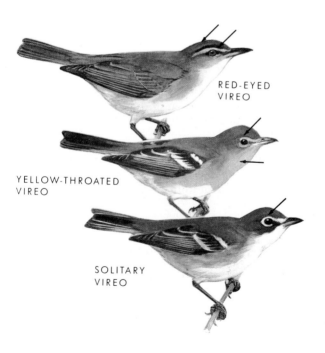

RED-EYED VIREO

YELLOW-THROATED VIREO

SOLITARY VIREO

WOOD WARBLERS. These birdlets, which make birding so exhilarating in spring and so exasperating in fall, are found only in the New World. Of 114 species, we have 52, plus 5 accidentals. Most winter in the tropics.

YELLOW-RUMPED WARBLER   5–6″

The *yellow rump, yellow patch* near the shoulder, and the note—a loud *check*—identify this species. Breeding males have a black breast patch; this is lost in winter, when they look more like the brownish females. This species comes in two forms: the "Myrtle" Warbler, which has a white throat and breeds in conifer forests from New England and the Maritimes across Canada to Alaska; and "Audubon's" Warbler, with a *yellow* throat. The latter form replaces the "Myrtle" in the conifer-clad mountains of the western U.S. Hardier than their kin, most Yellow-rumped Warblers winter within the U.S.

MAGNOLIA WARBLER   4¾″

The little Magnolia Warbler also has a yellow rump, but the *heavily striped* yellow underparts and *broad white band* midway across the tail are distinctive. The Magnolia Warbler breeds in the evergreen forests across Canada east of the Rockies, and in the higher hills and cooler forests of the northeastern U.S.

CANADA WARBLER   5–5¾″

The *necklace of short stripes* on its yellow breast marks this gray-backed warbler. It prefers forest undergrowth and sings a pleasant, lively burst of notes, irregularly arranged. Basically eastern, it breeds across wooded Canada east of the Rockies and south in the cooler forests of the northeastern tier of states and the higher Appalachians.

male

YELLOW-
RUMPED
WARBLER

"Myrtle"

female

("Audubon's")
western form

male

MAGNOLIA
WARBLER

CANADA
WARBLER

YELLOW WARBLER 5″

This is the only warbler that looks almost all-yellow; even the *tail spots* are *yellow*, a feature shared only by the dissimilar female Redstart (p. 98). The male has *rusty breast streaks*. From swamp edges, streams, and gardens we may hear the male's cheerful *weet weet weet weet tsee-tsee*, with variations. Widespread, the Yellow Warbler nests from treeline in Canada to the northern edge of the Gulf states and in much of the West.

COMMON YELLOWTHROAT 4½–5½″

The *black mask* is the field mark of the male; the female lacks this. This warbler's actions are almost wren-like as it bustles about the bushes in swampy places, singing a 3-syllabled *witchity-witchity-witchity-witch*. It breeds across Canada and in every one of the lower 48 states.

HOODED WARBLER 5½″

A *black hood* or *cowl* encircles the *yellow cheek* of the *male* Hooded Warbler; the female lacks this. Confined to the East, from the southern Great Lakes to the Gulf states, this warbler enlivens the woodland undergrowth and laurel thickets, singing its loud, whistled *weeta-weetee-o*.

YELLOW-BREASTED CHAT 6½–7½″

This unusually large warbler looks like an overgrown female Yellowthroat, but its *bill* is *larger* and its *tail* is *much longer*, giving it the shape of a Catbird. Its song is a series of clear, repeated whistles and phrases that suggest a slower-paced Mockingbird song. The Chat summers in tangles and thickets across the U.S., locally from the Canadian border to the Gulf states and Mexico.

YELLOW WARBLER

COMMON
YELLOWTHROAT

HOODED WARBLER

YELLOW-
BREASTED
CHAT

BLACK-THROATED GREEN WARBLER
4½–5″

The *bright yellow face* set off by the black throat is the trademark of this attractive warbler. Its song is a lisping *zoo zee zoo zoo zee,* or a variation of that theme. Favoring evergreens, it ranges across Canada east of the Rockies and south in the hemlock forests of the Appalachians.

TOWNSEND'S WARBLER   4½–5″

In western Canada and Alaska this pretty warbler replaces the Black-throated Green Warbler in the cool conifer forests, where it sings its wheezy *dzeer dzeer dzeer teetsy.* It differs from the Black-throated Green in having a *black cheek patch,* a black crown, and a *yellow belly.*

BLACK-THROATED GRAY WARBLER 4½–5″

This resident of the oaks and junipers of the western states is patterned almost identically like the more northern Townsend's Warbler (above), but *lacks the yellow* on its face and underparts. Its song is a buzzy chant, *zeedle zeedle zeedle zeet' che,* or a variation of it.

BLACK-AND-WHITE WARBLER   4½–5½″

Striped lengthwise with black and white, this warbler is unmistakable as it *creeps* along the trunks and branches, singing a thin *weesee weesee weesee weesee weesee weesee.* The Black-and-white Warbler is widespread in summer across southern Canada east of the Rockies and in the U.S. east of the Great Plains. It retreats in winter to the tropics, but some individuals may spend the winter in Florida and along the Gulf Coast.

BLACK-
THROATED
GREEN
WARBLER

TOWNSEND'S
WARBLER

BLACK-
THROATED
GRAY
WARBLER

BLACK-
AND-
WHITE
WARBLER

AMERICAN REDSTART 5″

Like an avian butterfly, the Redstart fans
its wings and tail—black and *orange* in the
male, olive-brown and *yellow* in the
female. It sings an insistent *zee zee zwee*
or *teetsa teetsa teetsa teetsa teet.* It
summers in second-growth woodlands
across Canada and the U.S. east of
the Rockies.

BLACKBURNIAN WARBLER 5″

The *flaming orange* of the throat and face
sets the Blackburnian Warbler apart. Its
song, *zip zip zip titi tseeeeee,* ends on
a slurred note so thin and high that many
ears cannot hear it. An inhabitant of
conifers in summer, it breeds in forests
across Canada east of the Rockies, as well
as in the cooler woods of the northeastern
states and the Appalachians.

CHESTNUT-SIDED WARBLER 4½–5½″

Well named, this little warbler can be
identified by its *chestnut sides.* Its bright
song sounds like *please please pleased ta
meet' cha.* Slashings and bushy pastures
coming into second growth are the habitat
of this warbler in summer, when it can
be found across southern Canada and in
the northern tier of states east of Minnesota
and southward in the Appalachians.

NORTHERN WATERTHRUSH 6″

Although the Waterthrush is really a
warbler, it has the look of a small thrush
that has been crossed with a sandpiper. It
often *walks* at the water's edge and *teeters*
like a Spotted Sandpiper. Its summer home
is Alaska, Canada, and the northern edge
of the U.S. east of the prairies. The similar
Louisiana Waterthrush (not shown) is
confined to the eastern states.

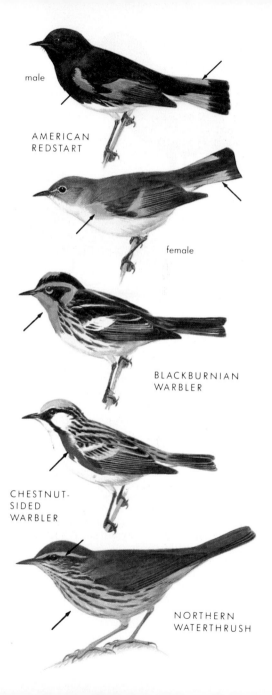

male

AMERICAN
REDSTART

female

BLACKBURNIAN
WARBLER

CHESTNUT-
SIDED
WARBLER

NORTHERN
WATERTHRUSH

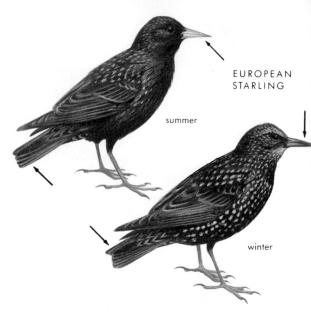

EUROPEAN
STARLING

summer

winter

STARLINGS are an Old World family with
103 species, many of which resemble our
blackbirds (icterids). Two species, the
Starling and the Crested Myna, have been
introduced into North America.

EUROPEAN STARLING   7½–8½″

This is the *short-tailed* "blackbird" so
common around cities, parks, and farms. It
gangs up in great flocks, roosting commun-
ally in winter under bridges and on
buildings. In summer it sports a *yellow
bill* and its black plumage is somewhat
glossed with iridescence. In winter the bird
is *heavily speckled* and the bill becomes
dark. Its many notes include a characteristic
whistled *whooee*. Versatile, it sometimes
mimics other birds.

BLACKBIRDS, ETC. (opposite). A varied
New World family known as Icteridae. Some
icterids are black, others highly colored.
There are 88 species; we have 20, plus
3 strays.

RED-WINGED BLACKBIRD   7–9½″

The male Red-wing with its *red epaulettes* is unmistakable, but its mate is a dull brown, heavily streaked bird, recognizable as a blackbird by its sharp bill and flat profile. The male's song, a typical sound of spring, is a gurgling *konk-la-ree* or *o-ka-lay.* There is scarcely a county in the lower 48 states where it does not nest, depending on the availability of marshes, swamps, and hayfields. It is widespread across southern Canada as well, withdrawing in winter.

YELLOW-HEADED BLACKBIRD   8–11″

The Yellow-head, a colonial blackbird of the western half of the continent, often shares the marshes with the Red-wing. Males have *orange-yellow heads* and white wing patches; females are brown, with yellow only on the throat and chest. The Yellow-head utters, with effort, rasping notes that sound like rusty hinges.

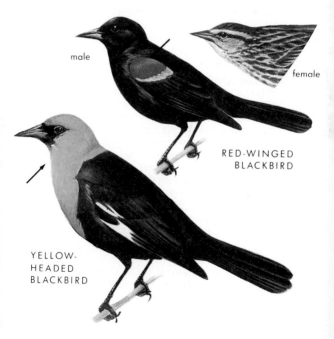

male

female

RED-WINGED
BLACKBIRD

YELLOW-
HEADED
BLACKBIRD

BROWN-HEADED COWBIRD · 7″

A small, black-bodied bird, a bit larger than a House Sparrow, with a *brown head* and a rather finchlike bill. Females are nondescript gray with the finchlike bill. A brood parasite, the Cowbird lays its eggs in the nests of other birds. It breeds from coast to coast and from the southern provinces of Canada to northern Mexico and the Gulf states, but not in Florida, where it occurs only during winter.

BREWER'S BLACKBIRD · 9″

In the West, Brewer's Blackbird replaces the familiar Grackle of the East as a bird of towns and farms. Males are all-black with a *shorter tail* than the Grackle's and only a slight iridescent gloss. Females are brownish gray with dark eyes. Brewer's Blackbird has recently extended its range eastward to the western Great Lakes states. Its counterpart in the East, the **Rusty Blackbird** (not shown), spends the summer in the wooded swamps and muskegs of Canada and seeks river groves and timbered swamps during its winter sojourn in the eastern U.S.

COMMON GRACKLE · 11–13½″

A large, highly iridescent blackbird with a *wedge-shaped or keel-shaped tail.* This familiar bird walks sedately on the lawns of suburban towns, in the fields, and along the margins of streams. It is widespread east of the Rockies from southern Canada to the Gulf states, often wintering as far north as the southern Great Lakes and southern New England. A larger, longer-tailed bird, the **Boat-tailed Grackle** (not shown), resides in the coastal marshes of the southeastern states.

BROWN-HEADED
COWBIRD

BREWER'S
BLACKBIRD

COMMON GRACKLE

male

female

ORIOLES. Smaller and slimmer than a Robin, these brightly colored birds belong to the family Icteridae and thus are closely related to the blackbirds. Although orioles are predominantly tropical, 7 species nest in the U.S.; 2 are accidental strays.

NORTHERN ("BALTIMORE") ORIOLE   7–8″
Of the 7 orioles normally found in the U.S., the Northern Oriole is the best known. The eastern form, widespread east of the plains, is known as the "Baltimore" Oriole and is recognized by its *flame-orange* and black pattern and *black head.* Its piping whistles accompany the days of early summer. Females are dull greenish above, yellowish below. The purse-like nest is suspended from a branch tip, often overhanging a city street.

NORTHERN ("BULLOCK'S") ORIOLE   7–8″
"Bullock's" Oriole, the western form of the Northern Oriole, differs from the "Baltimore" form by its *orange cheeks* and *large white wing patches.* This, the most widespread oriole in the West, overlaps the range of the "Baltimore" on the plains. The two were formerly regarded as separate species, but since they may hybridize where they meet on the Great Plains, taxonomists have lumped them.

ORCHARD ORIOLE   6–7″
Unlike other orioles, which are either bright orange or yellow, the Orchard Oriole is dark-hued. Its rump and underparts are *deep rich chestnut.* Females are somewhat greener-looking than female Northern Orioles. The song is a fast-moving outburst with some piping notes. A bird of the East, it ranges widely across the U.S. from the Atlantic to the plains, favoring orchards, farms, and towns.

NORTHERN ("BALTIMORE") ORIOLE

NORTHERN ("BULLOCK'S") ORIOLE

ORCHARD ORIOLE

BOBOLINK

MEADOWLARK

Meadow Birds (More Icterids)

BOBOLINK   6–8″

A bird of the fields of northern U.S. and
southern Canada. The spring male is our
only songbird that is *black below* and
largely white above. The song, delivered in
hovering display flight, is ecstatic and
bubbling. Before migrating to the winter
home in Argentina, the male takes off his
dress-suit and assumes a tawny, sparrow-
like plumage not unlike that of the female.

MEADOWLARK (2 species)   9″

In grassy fields, a chunky brown bird flies
away, showing a conspicuous patch of
white on each side of its short broad tail.
Perched on a post, it presents a *bold black
V* on a yellow breast. The Western
Meadowlark, shown here, is very similar to
the Eastern Meadowlark but sings a
totally different song: a tumble of gurgling

flutelike notes, quite unlike the clear slurred whistles of the Eastern Meadowlark. The ranges of the two birds overlap in the prairie states.

TRUE LARKS. These streaked, brown, terrestrial birds number 75 species in the Old World, of which only 1, the Horned Lark, is also found widely in North America. Another, the Skylark of Europe, has been introduced on Vancouver Island.

HORNED LARK   7–8″

On open ground—prairies, fields, golf courses, and shores—at the right season you will find this "shorelark." Its *facial pattern, "horns,"* and *black breast patch* identify it. Its song, high and tinkling, may be given from the ground or high in the air, in skylark fashion. Widespread in suitable environments from the high Arctic to Mexico, it is absent only from the southeastern states.

HORNED LARK

TANAGERS—"THE BRIGHT ONES." A

New World family of 190 species; most live in the tropics. In the U.S. we have 4; the males of 3 are shown here. The females are olive-backed, yellow-breasted birds that may be confused with female orioles.

SCARLET TANAGER 7″

The *flaming scarlet body* of the male contrasts with its *jet-black wings and tail.* The Scarlet Tanager sings in short phrases—like a Robin with a sore throat. A woodland bird, it favors oaks from southern Canada throughout the northeastern and central states. When it departs for the tropics in early autumn, the male assumes a plumage much like that of its dull greenish mate.

SUMMER TANAGER 7–7¾″

This rose-red bird of the southern woods and oak groves is *red all over,* lacking the black wings of the Scarlet Tanager. Its song suggests that of a Robin, but its note—a staccato *pik-i-tuk*—is distinctive. The Summer Tanager breeds in the central and southeastern states, from the Atlantic Coast west to Oklahoma and Texas.

WESTERN TANAGER 7″

The Western Tanager, with its *red face* and pattern of yellow and black, is one of the most colorful birds in the West. The 2 wing bars also separate it from the other tanagers, but may cause confusion with female orioles. The song resembles that of the Scarlet Tanager. It lives in open pine or mixed forests from southwestern Canada nearly to the Mexican border.

SCARLET
TANAGER

female

male

SUMMER
TANAGER

WESTERN
TANAGER

GROSBEAKS, FINCHES, ETC. These birds
have short, seed-cracking bills: large and
thick in the grosbeaks, smaller and
more canarylike in the finches, sparrows,
and buntings. There are 245 species in the
world; 70 in North America, plus 4 strays.

NORTHERN CARDINAL   7½–9″
The Cardinal, all-red, with its *pointed
crest, black face*, and *triangular red bill*, is
a garden favorite. Its song is a series of
clear slurred whistles—*what cheer cheer
cheer*, etc., or *birdy birdy birdy birdy
birdy*. The female is buffy-brown with some
red in the wings, tail, crest, and bill.
Widespread south of Canada and west to
the plains, the Cardinal has extended
its range north in recent years, mainly
because so many people are putting out
sunflower seeds.

ROSE-BREASTED GROSBEAK   7–8½″
This handsome bird, with its *big pale bill*
and *rose-red breast patch*, is a woodland
bird, more often heard than seen. Its fluid
song suggests a Robin "that has taken
voice lessons." The call note, a metallic *eek*,
is distinctive. The female looks like an
overgrown female Purple Finch (see p. 112),
except for the *large* "grosbeak" *bill*. A
summer resident of southern Canada and
the northern U.S. east of the plains, this
grosbeak winters in the tropics.

BLACK-HEADED GROSBEAK   6½–7¾″
This grosbeak, with its *dull orange-brown*
breast and rump, replaces the Rose-
breast in western North America. The
female is similar to the female Rose-breast
but is much *less streaked* on the breast,
which is *washed with tawny*. The song of
the male is much like that of the Rose-
breast.

NORTHERN
CARDINAL

female

male

ROSE-
BREASTED
GROSBEAK

male

female

male

female

BLACK-
HEADED
GROSBEAK

RED FINCHES

PURPLE FINCH 5½–6″

The Purple Finch is a frequent visitor to feeding trays, where it gorges on sunflower seeds. The male looks like a sparrow *dipped in raspberry* juice. The female suggests a heavily striped sparrow with a dark ear patch and a *heavy jaw stripe.* The song is a fast lively warble, heard at its best on nesting territory among small evergreens in Canada, the northeastern U.S., and the Pacific states. The note, a metallic *tick,* is quite unlike any note of the House Finch. Some Purple Finches may winter where they nest; others may wander as far south as the Gulf of Mexico.

HOUSE FINCH 5–5¾″

Formerly a bird of the West, the House Finch was accidentally introduced near New York City in the 1940s and has spread over the Eastern Seaboard states, where it now shares the feeders with the Purple Finch. It is more adapted to cities, towns, and suburbs than that bird and may partially displace the House Sparrow. Males are separated from Purple Finches by their smaller-headed look and *striped sides.* Females lack the strong head pattern of the female Purple Finch.

RED CROSSBILL 5¼–6½″

The Red Crossbill, a *deep red* bird with dark wings, is not likely to be spotted by beginning birders except in Maine or in evergreen-clad parts of Canada and the West. The specialized *bill with crossed tips* is used to extract seeds from evergreen cones. Periodically, when the cone crop fails, Crossbills move out and may even turn up at feeding trays. The less common **White-winged Crossbill** (not shown) has conspicuous *white wing bars.*

male

female

PURPLE
FINCH

male

female

HOUSE
FINCH

female

male

RED
CROSSBILL

INDIGO BUNTING 5½"

The male Indigo Bunting, a small goldfinch-sized bird, is *deep blue all over* (but see the Mountain Bluebird on p. 89). The female is as plain as a bird can be—brown with no really distinctive marks. The song of the male, delivered from a wire or bush top in brushy country, is lively, high, and strident, with the notes usually paired. Summering east of the plains, from southern Canada to the Gulf states, the Indigo Bunting winters mainly in the West Indies, Mexico, and Central America, but a few may linger in southern Florida. In winter the male assumes a dull plumage much like that of the female.

LAZULI BUNTING 5–5½"

The Lazuli Bunting replaces the Indigo Bunting in the West. It is colored vaguely like the Bluebird (p. 89), but notice the stubby bill and *white wing bars.* Its song is similar to that of the Indigo Bunting. Where the two species meet on the plains they sometimes hybridize. The Lazuli Bunting, common in mountain country, prefers dry brushy slopes and streamsides, sagebrush, briars, and recently burned areas.

PAINTED BUNTING 5¼"

No American bird is gaudier than the male Painted Bunting, with its *purple head, green back,* and *scarlet* underparts. In contrast, its mate is a plain little *greenish* finch. The male's song is a bright, pleasant warble. The Painted Bunting has a restricted range along the Atlantic Coast from North Carolina to Florida, but is also found from the lower Mississippi River to southern New Mexico.

INDIGO
BUNTING

LAZULI
BUNTING

PAINTED
BUNTING

AMERICAN GOLDFINCH 5″

In summer this tiny finch with the roller-coaster flight and canarylike song is readily identified by its *black wings*, tail, and cap. In winter, the male loses the bright yellow and looks much like the drab female. Goldfinches are feeding-tray addicts and are especially fond of thistle seed. Found at one season or another over much of North America, they are partially migratory.

PINE SISKIN 4½–5″

This tiny finch of the evergreen forests that stretch across Canada and the highlands of the West is a nomad, wandering erratically in winter as far as the Gulf of Mexico. During invasion years it might come to the feeding tray. The size and shape of a Goldfinch, it is very *heavily streaked*, usually with a *bar of yellow* on the wing and a *touch of yellow* at the base of the tail. Its voice, as it flies overhead, is a light *tit-i-tit* or a buzzy *shreeeee*.

EVENING GROSBEAK 8″

The chunky, *big-billed*, short-tailed males are deep yellow with *black and white wings*; females are grayish, with enough of the males' features to be recognized. After nesting in evergreen forests of Canada and the mountainous West, Evening Grosbeaks wander widely in winter. Flocks of these large, boldly patterned winter finches can be attracted to the window tray by putting out sunflower seeds. Because of this subsidy of food, Evening Grosbeaks have extended their range eastward and may nest even in northern New England and the Maritimes.

AMERICAN GOLDFINCH

male summer

female

PINE SISKIN

EVENING GROSBEAK

SNOW BUNTING   6–7¼″

During summer, when the Snow Bunting sings to the Inuit on the tundra, the males are black-backed, quite unlike the winter bird shown here. Although some birds may look quite brown as they glean the weed seeds in open fields, their flashing *white wing patches* identify them. A resident of the Arctic in summer, the Snow Bunting travels widely in winter. Its flocks give life to the snow-covered prairies, fields, and shores as far south at times as the central states.

NORTHERN JUNCO   5½–6¾″

These hooded, sparrow-shaped birds are best known in winter, when they patronize the ground feeder. The *white outer tail feathers* are twitched as the bird feeds or are flashed when it flies away. Eastern birds ("Slate-colored Juncos") have *slaty* backs and gray flanks; western birds ("Oregon Juncos") have *rusty* backs and flanks. The song is a loose, musical trill on one pitch. This summer resident of the cooler forests of the North winters throughout the U.S.

RUFOUS-SIDED TOWHEE   7–8½″

The Towhee is a bird of open woods and undergrowth, where it rummages among the dead leaves, kicking them with both feet at once. The hooded effect—black in the male, brown in the female—as well as the *rusty sides* and flashing white tail patches are diagnostic. In the West, Towhees show additional white spots on the back and wings. The eastern bird sings *drink-your-teeeee.* The note is a loud *che-wink!* Towhees breed from southern Canada to the Gulf states, but withdraw from the snowbound north in winter.

SNOW
BUNTING

"Slate-colored"

NORTHERN
JUNCO

"Oregon"

male

RUFOUS-
SIDED
TOWHEE

female

spotted
western
race

Sparrows (with rusty caps)

CHIPPING SPARROW — 5¼"

The *rusty cap* and the *white eyebrow*
identify the little Chippy, which is adapted
to suburban gardens and farms over
much of the U.S. and Canada. Its *unmarked*
breast sets it apart from its neighbor, the
Song Sparrow (p. 122), which has a
streaked breast. The song of the Chipping
Sparrow is a chipping rattle on one pitch.

AMERICAN TREE SPARROW — 6–6½"

The single *black "stickpin"* on its otherwise
unmarked breast marks the "Winter
Chippy," which like its smaller relative has
a rusty cap. A resident of arctic scrub in
summer, the Tree Sparrow fills the void left
in winter by the Chippy, which has
abandoned the northern states. This
sparrow forages in little parties along
brushy roadsides, weedy edges, and
marshes.

FIELD SPARROW — 5"

This rusty-capped sparrow can be separated
from its look-alikes by its *pink bill*. Its
song is a series of sweet slurred notes
speeding into a trill which may ascend,
descend, or stay on the same pitch. Despite
its name, this sparrow inhabits brushy
pastures and scrubby edges, not fields.
Widespread east of the Rockies, it withdraws
in winter from the snowbound north.

SWAMP SPARROW — 5–5¾"

This rather stout, dark sparrow of brushy
swamps and cattail marshes exhibits a
rusty cap and a *whitish throat*. Its song is
a loose trill, suggesting the song of the
Chippy but slower and sweeter. Summering
across Canada east of the Rockies and in
the northeastern block of states, it winters
from Lake Erie and southern New England
to the Gulf states.

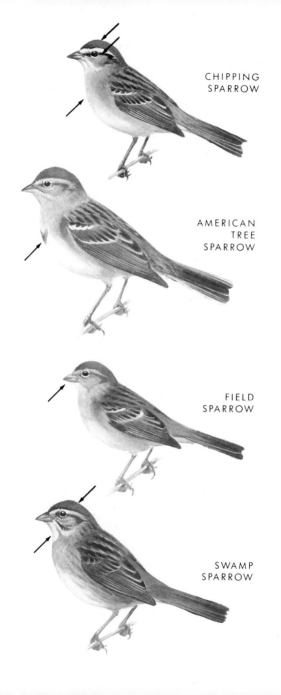

CHIPPING SPARROW

AMERICAN TREE SPARROW

FIELD SPARROW

SWAMP SPARROW

Sparrows (with striped breasts)

SONG SPARROW   5–6½″

In this familiar sparrow the streaks on the
breast coalesce into a *large central spot.*
The tail, not notched as in some of the
other streaked sparrows, dips up and down
as the bird flies from bush to bush. The
Song Sparrow is a harbinger of spring in
the North, where it often appears before the
last snows have melted. Its musical song
usually starts off by repeating a bright
introductory note 3 or 4 times, *sweet
sweet sweet,* etc.

SAVANNAH SPARROW   4½–5¾″

Superficially like a Song Sparrow with a
shorter, *notched* tail, the Savannah
Sparrow is more of an open country bird. It
usually lacks the large central breast spot.
The song is a dreamy, lisping *tsit-tsit-tsit-
tseeeee-tsaaay.* It is widespread across
North America from the Arctic Slope south
throughout much of the West and the
northeastern block of states. In winter most
Savannah Sparrows avoid the icy fields
and are to be found in the snow-free
southern states.

FOX SPARROW   6¾–7½″

This big, ground-loving sparrow of the
Canadian north woods and western mountains can be identified by its heavily
blotched breast and *rusty tail.* When
foraging among dead leaves it kicks with
both feet at once. The Hermit Thrush also
has a reddish tail, but its slender beak is
quite unlike the thick bill of the Fox
Sparrow. Some Fox Sparrows move south
in winter as far as the Gulf states and may
patronize the ground feeder. The bird
shown here is the eastern form. Western
birds come in various shades of brown and
gray but have the rusty tail.

SONG SPARROW

SAVANNAH SPARROW

FOX SPARROW

WHITE-THROATED SPARROW 6½–7″

The White-throat is the voice of the cool forests that extend across Canada from the Maritimes to the Rockies. The *white throat patch* and *striped crown* (black and white, or brown and tan) identify it. Its pensive song, 1 or 2 clear whistles followed by 3 quavering notes on a different pitch, is easily imitated. Wintering in woods and thickets throughout the East, the White-throat is often a patron of the ground feeder.

WHITE-CROWNED SPARROW 6½–7½″

The White-crowned Sparrow, with its clear gray breast and puffy *striped crown*, differs from its more eastern relative, the White-throat, in having a *pink bill*. It also lacks the well-defined white throat patch. Its plaintive song varies—usually one or more clear whistles followed by husky, trilled whistles. Breeding across boreal Canada and southward through western North America, it is known in the East only in migration and winter.

GOLDEN-CROWNED SPARROW 6–7″

This sparrow of western Canada and Alaska winters in the Pacific states. There it often associates with the White-crowned Sparrow, from which it may be distinguished by its *golden crown.*

DICKCISSEL 6–7″

The stronghold of the Dickcissel is the Midwest, where it sings its staccato *Dick-ciss-ciss-ciss* from the fields and prairies. The male with his *black bib* and *yellow chest* suggests a miniature Meadowlark, but the female looks more like a female House Sparrow (except for the rusty shoulders and *touch of yellow* on the chest).

tan-striped form

WHITE-
THROATED
SPARROW

white-striped form

immature

WHITE-
CROWNED
SPARROW

adult

GOLDEN-
CROWNED
SPARROW

DICKCISSEL

HOUSE
SPARROW

female

male

WEAVER FINCHES. This large group of Old
World birds numbers 263 species. The
sparrow-weavers, the group to which the
House Sparrow belongs, number 35
species, 2 of which have been introduced in
the U.S. They are not related to our native
sparrows.

HOUSE SPARROW   6″

Everyone knows the House Sparrow. Birds
living in polluted city air are often quite
sooty, bearing little resemblance to the
clean country male shown here, with its
black throat and chest patch, light cheeks,
and *chestnut nape.* Females have a plain
dingy breast and a dull eye-stripe. The
House Sparrow, introduced from Europe in
1850, was released in Brooklyn. From
there it spread rapidly until it now lives
from coast to coast wherever there are
towns or farms, and from central Canada
southward through the United States to
Central and South America, where it is still
pioneering.

Index

Color illustrations of the birds generally appear on the page facing the text. To avoid duplication, the illustrations are not indexed separately.